An Entangled Bank

An Entangled Bank

The Origins of Ecosystem Ecology

Joel B. Hagen

RUTGERS UNIVERSITY PRESS
New Brunswick, New Jersey

Library of Congress Cataloging-in-Publication Data

Hagen, Joel Bartholemew.
An entangled bank : the origins of ecosystem ecology / Joel B. Hagen.
p. cm.
Includes bibliographical references and index.
ISBN 0-8135-1823-7 (cloth) — ISBN 0-8135-1824-5 (paper)
1. Ecology—History. 2. Ecology—United States—History.
I. Title.
QH540.8.H34 1992
574.5'09—dc20 91-41840
CIP

British Cataloging-in-Publication information available

Manufactured in the United States of America

To

NORM FORD and PAUL FARBER,

Teachers and Friends

Contents

Preface

I BEGAN WORKING on this book in 1987 while I was a visiting assistant professor at the University of Maryland. I wish to thank the members of the Department of History and the Committee on History and Philosophy of Science for supporting and encouraging me during the early stages of research. I am particularly grateful to the reading group of Lindley Darden, Pamela Henson, and Joe Cain who commented on an early draft of the book.

During the course of my research I had the opportunity to interview several biologists about their work or about historical figures with whom they were acquainted. Others generously responded to my questions in writing. These include F. Herbert Bormann, Jim Collins, Maxwell Dunbar, W. T. Edmondson, Doug Gill, Frank Golley, Jack Harley, William Hiesey, G. Evelyn Hutchinson, David Keck, Eugene Odum, Howard Odum, Bernard Patten, Charles Reif, Peter Rich, Paul Richards, John Teal, Saran Twombly, and Bill Winner.

I greatly appreciate the opportunity to study unpublished letters and manuscripts at the following institutions: American Heritage Center, University of Wyoming; University of Minnesota Archives; Smithsonian Institution Archives; Magdalen College Archives, Oxford University; Department of Botany, Cambridge University; Department of Botany, Oxford University; Carnegie Institution of Washington Archives; Sterling Library, Yale University; Office of the Registrar, Park College. I wish to acknowledge the generous assistance I received at all these institutions. I also wish to thank the librarians at Radford University, particularly Bud Bennett, for their help. I am most grateful to Nancy Slack for helping to arrange access to the G. Evelyn Hutchinson papers during a visit to Yale University. She has been an excellent source of information on Hutchinson and his students. Tobey Appel kindly allowed me to read parts of her unpublished history of biology at the National Science Foundation. She patiently answered many of my questions about post–World War II funding of science. Peter Taylor provided me with useful information about the work of Howard Odum. Although we disagree on some points of interpretation, his insistence on the importance of Odum's work has influenced my thinking on the history of modern ecology. Jim Collins read the entire manuscript, corrected many errors, and made numerous helpful suggestions. I particularly appreciate the time that he has so generously spent on the telephone discussing my research as it progressed.

This research was supported by grants from the National Endowment for the Humanities, the Smithsonian Institution, and the Radford University Foundation.

Finally, I wish to thank Susan, Kirsten, and Christopher for putting up with a historian in the family.

An Entangled Bank

1

An Entangled Bank

Battle within battle must ever be recurring with varying success; and yet in the long-run the forces are so nicely balanced, that the face of nature remains uniform for long periods of time.

—CHARLES DARWIN, *On the Origin of Species*

CHARLES DARWIN presented an ambiguous picture of nature in his greatest work, *On the Origin of Species.* Nature was a battlefield on which individuals ceaselessly struggled in the "war of nature," but it was also a stable complex of interacting parts. Indeed, a recurring theme in the book is the "entangled bank" covered with diverse plants and animals interacting according to definite laws of nature similar to those governing the movement of the planets.[1] Darwin marveled at nature's "web of complex relations." Take for example, the close interactions among flowering plants, humble-bees, mice, and cats near Down House. According to Darwin, experience showed that red clover almost totally depended upon humble-bees for pollination; other bees did not visit the clover because they could not reach the nectar in the narrow, tubular flowers. The population of humble-bees was regulated by field mice that destroyed the bees' nests. The number of mice depended on the number of cats in the neighborhood. Thus, Darwin concluded, the population of clover might well depend indirectly on the population of cats.[2]

Darwin's example provides an excellent illustration of basic ecological principles. Species do not exist completely independently, but they often form interacting groups. Regulation of one population by another may be indirect. Sometimes a single species may have a pervasive influence on several other members of the web. For the modern reader all this is immediately obvious in Darwin's writing. Yet despite his confidence that the struggle for existence could explain natural order, Darwin did not rigorously do so in 1859. His discussions of nature's "entangled bank" were short, literary passages interspersed in his more technical discussion of speciation and the evolution of adaptations. Only during the twentieth century did biologists propose general theories to explain the type of observations that Darwin made near Down House. This became the intellectual domain of community and ecosystem ecology.

Darwin's two views of the living world—machinelike stability and chaotic warfare—appear anomalous. But were they? Historians disagree on this matter. In his history of ecological ideas, *Nature's Economy,* Donald Worster emphasizes the inherent contradiction between these two views.[3] In fact, he claims that this intellectual dichotomy reflected a fundamental division in Darwin's psyche. His pastoral existence at Down House and the competitive professional life of London represented psychological poles analogous to the entangled bank and the battlefield of nature. Both intellectually and psychologically he struggled with these polarities, but in the end they remained unreconciled. According to Worster, Darwin may have been a reluctant revolutionary, trying to temporize the idea of the struggle for existence, but violent encounter remained the dominant theme in both his evolutionary writings and psychological character.[4]

Quite a different interpretation is presented by Edward Manier. According to Manier, Darwin's concept of struggle for existence was a deliberate choice, a compromise between Thomas Hobbes's war of nature and Charles Lyell's idea of nature in a steady-state.[5] For Darwin, the struggle for existence was an extremely flexible concept that included not only face-to-face competition, but also differential reproduction, parasitism, mutualism, and adaptation to the physical environment.[6] The indeterminacy implied by natural selection fit somewhat uncomfortably with the Newtonian clockwork universe so central to the Victorian world view, and, in the end, evolution proved profoundly subversive to Victorian beliefs in stability, natural order, and progress.[7] But this was not obvious even to Darwin, who, though tending toward a view of natural laws as statistical summaries of phenomena, never completely broke with the more traditional notion of deterministic laws of nature.[8]

Several prominent ecologists have recently argued for a historical interpretation curiously similar to Worster's, with its emphasis upon the contradictory character of Darwin's composite view of nature.[9] These ecologists, all critics of the idea that nature is in equilibrium, have drawn sharp distinctions between determinism and indeterminism, stability and instability, stasis and change. These dichotomies, so it is claimed, reflect antithetical intellectual positions deeply rooted in different cultural matrices. Historically, these ecologists contend, their discipline is grounded in a dogmatic commitment to the idea that nature is in equilibrium; only recently have ecologists recognized that the living world is characterized by pervasive disturbance and instability.

Whatever scientific merits nonequilibrium ecology may have, the historical claims of its proponents can be challenged on two grounds. First, like Darwin, other nineteenth-century proto-ecologists sought an intermediate position, one that could account for both stability and instability in the natural world. Second, it appears that the transition from the Victorian clockwork universe to a more indeterminate world of instability and change produced a creative tension in biology. Far from dogmatic adherence to naive notions of equilibrium, late nineteenth-century biologists forged a set of flexible concepts for dealing with the evolutionary complexities of the natural world. These concepts were inherited by ecologists when the new discipline began to form during the early decades of the twentieth century.

The Social Organism

Although Darwin's work remained largely within the conceptual framework of nineteenth-century natural history, natural selection suggested strikingly new ways of looking at life, in general. For example, in an essay review of Ernst Haeckel's *The Natural History of Creation,* Thomas Henry Huxley speculated that natural selection might be extended into the realm of physiology. According to Huxley,

> It is a probable hypothesis, that what the world is to organisms in general, each organism is to the molecules of which it is composed. Multitudes of these, having diverse tendencies, are competing with one another for opportunity to exist and multiply; and the organism, as a whole, is as much the product of the molecules which are victorious as the Fauna, or Flora, of a country is the product of the victorious organic beings in it.[10]

Physiologically, Huxley believed, both heredity and adaptation could be explained in terms of the differential multiplication and survival of organic molecules.

Viewing an individual organism, or even a cell, as a kind of population, community, or ecosystem composed of interacting microscopic parts is an old and amazingly resilient idea.[11] During the late nineteenth century, Huxley was not the only one to envision the struggle for existence occurring within the apparently stable, multicellular organism. An even broader claim was made by Herbert Spencer. Although Huxley extended the notion of natural selection to the microcosm of the individual organism, Spencer's evolutionary philosophy employed the struggle for existence as a general law of nature acting at all levels of organization. For example, in his early essay, "The Social Organism," Spencer applied the struggle for existence to both the physiological microcosm and the social macrocosm.[12] As the title of the essay suggests, Spencer saw a close parallel between the physiological body and the body politic.

Despite the fact that Spencer's 1860 essay dealt specifically with human societies, it is particularly important to consider within the context of the history of ecology. The essay is perhaps the clearest and certainly the most concise statement of Spencer's organic analogy, a concept that was borrowed by a diverse group of late nineteenth- and early twentieth-century intellectuals. Historians have disagreed sharply over the extent of Spencer's influence, particularly in America. But, even those historians who have challenged the popular portrait of Spencer as a kind of late nineteenth-century American folk hero have acknowledged that his ideas, though often misunderstood and frequently modified, had a significant impact on American social thought.[13] Historians of biology have argued that directly or indirectly Spencer influenced the first generation of American ecologists.[14] Specifically, the idea that a group of plants and animals, or biological community, can be thought of as a kind of organism became an important element in the conceptual framework of ecology.

Spencer's programmatic goal was to explain society in biological, and ultimately physical, terms.[15] What emerged from "The Social Organism" was a mechanical/organic model that, although somewhat incongruous, was widely copied by later thinkers. Unlike many later thinkers, however, Spencer quite carefully defined exactly what he meant by "organism." For Spencer, organic entities, whether individual organisms or human societies, shared a number of distinguishing characteristics. An organism increased its mass through an orderly process of *growth*. Unlike the type of growth characteristic of non-organismal objects such as crystals, organic growth entailed *differentiation*, an increase in complexity and division of labor. Finally, organisms exhibited important *part/whole relationships*. The parts of an organism were interdependent; ultimately the operation of a single part of the organism depended upon the smooth operation of other parts. The whole organism had a more prolonged life than had its parts. The organism persisted as several generations of individual parts arose, grew, did their work, reproduced, and died.

Spencer was also careful to discuss possible differences between individual organisms and social organisms. For example, critics might argue that while individual organisms have definite boundaries, societies rarely have a well-defined external form. But this was only a problem if one compared societies to higher animals; if one were to consider lower plants and animals, Spencer argued, one would find the same indefinite boundaries encountered in human societies. However, Spencer considered one difference between individual organisms and social organisms critical. Although physiologically one could speak of the parts of an individual organism being subordinate to the whole, Spencer's commitment to laissez faire precluded such a relationship in human society. In society the welfare of the individual could never be sacrificed for the welfare of the whole. This difference apparently was not sufficiently troubling for Spencer to reject the organic analogy, but some others did consider it a serious problem.

Spencer's organic analogy provided later intellectuals with concepts and a language for discussing social groups in terms of development, part/whole relationships, interdependence, and integration. But the essay illustrated some inherent problems with comparing communities—whether human or biological—with organisms. Indeed, the very brevity of the essay highlighted the inconsistencies in Spencer's thought. A glaring weakness in Spencer's argument is the rather naive anthropomorphism of his organic analogy. Ironically, he himself was aware of this problem. Early in the essay Spencer criticized Plato and Hobbes for drawing analogies between social structures and organs in the human body: "Both thinkers assume that the organization of a society is comparable, not simply to the organization of a living body in general, but to the organization of the human body in particular. There is no warrant whatever for assuming this."[16] However, his conviction that both social and biological evolution followed a progressive, linear path from simple to complex, undermined this cautionary note, and Spencer proceeded to draw parallels between Victorian society, the apex of social evolution, and the vertebrate (if not the human) body. Using the type of mechanical-organic images commonly employed by Victorian writers, Spencer compared telegraph lines to nerves, railroad systems to arteries, and currency to red blood cells.[17] This tendency to equate social groups not just with some type of organism, but specifically with the highly integrated vertebrate organism has been an inherent problem with the organic analogy. During the twentieth century, biologists—both adherents and critics of the analogy—have all too willingly assumed that if biological communities or ecosystems are like organisms, then perforce they must have structures analogous to nervous systems or endocrine glands.[18]

A deeper problem in Spencer's essay, and one shared with Darwin's "entangled bank" passage, is the ambiguous relationship between competition and social stability. Although Spencer supported the socioeconomic status quo, the social theory that he advocated was, as Sidney Fine suggested, "but one step removed from anarchism."[19] Written early in his career, "The Social Organism" reflects Spencer's optimism that unregulated competition produces social stability. Elaborating on the struggle-of-the-parts theme discussed earlier, Spencer compared competition in society to the physiological competition that supposedly occurred within the body. According to Spencer, "different parts of the social organism, like the different parts of an individual organism, compete for nutriment; and severally obtain more or less of it according as they are discharging more or less duty."[20] Just as during exercise blood is diverted from digestive organs to muscles, Spencer believed that certain economic activities such as railroad expansion would temporarily divert capital from other less active industries. For the individual human within society such competition might lead to bankruptcy, and for the individual parts of the body competition might lead to atrophy; but the social consequences of competition, in both cases, were stability and progress.

It is ironic that Spencer used the expansion of railroads to demonstrate how laissez faire leads to social stability and progress. In the United States, where Spencer was so widely admired, the expansion of railroads during the post–Civil War era resulted in social strife and contributed to the economic depression of 1873–1878.[21] Contrary to Spencer's vision of unregulated competition among independent individuals, this expansion eventually resulted in the growth of industrial and governmental bureaucracy. Fearing bankruptcy, both railroad managers and investors sought to minimize ruinous competition and increasingly turned to consolidation, integration, and cooperation.[22]

The ambiguity of portraying the well-regulated social organism as a site of unregulated competition was not lost on Spencer's critics. Huxley, who had no difficulty accepting competition among molecules in the body, balked at Spencer's attempt to explain social stability in the organic body politic in terms of laissez faire.[23] The organic analogy could not be used to justify unregulated competition among individuals in society, Huxley argued, "if the analogy of the body politic with the body physiological counts for anything, it seems to me to be in favour of a much larger amount of governmental interference than exists at present."[24]

For this reason, Huxley was not particularly drawn to Spencer's organic analogy, but a diverse group of philosophers, political scientists, sociologists, and historians later embraced both the organic analogy and the belief that this analogy justified a greater regulatory role for government. For example, the historian and social critic Charles Beard, whose work exemplified the newer form of cultural organicism so influential by the end of the century, cited both Darwin and Spencer. Beard argued, "It is generally recognised that society is more than a mere aggregate of individuals; that the individual is not only a sharer in the life of the organism, but is also capable of modifying by his inter-social activities its structure, function and lines of development."[25] Rejecting Spencerian individualism, Beard called instead for a rational, planned economy. Beard and other observers of human nature could use teleology to explain stability in the social organism, but this option was less acceptable to biologists. Thus the question remained: If the biological community were to be compared to an organism, could the struggle for existence explain the apparent stability of this organic entity? If so, exactly how did this occur? If not, what other mechanisms might be involved in maintaining stability?

The Lake as a Social Microcosm

The themes developed in Spencer's "Social Organism" were elaborated in ecological form in a classic essay written by Stephen A. Forbes. First published in 1887, "The Lake as a Microcosm" is generally recognized as one of the first statements of the ecological concept of the biological community.[26] So popular was this essay that it was reprinted in 1925, and it continued to be read and commented upon by ecologists for several decades thereafter. Writing during a period of professionalization and increasing specialization, Forbes was a transitional figure in the history of modern biology. Largely self-educated, he was one of the last great naturalists whose interests spanned the gamut of topics in traditional natural history: botany, entomology, ichthyology, and ornithology. At the same time, his seminal writings helped to define the newly emerging specialties of ecology and limnology.[27]

Born in Illinois to a poor farming family, Forbes's early education—one year at Beloit Academy in Wisconsin—was interrupted by military service during the Civil War. After serving in the Illinois cavalry, Forbes entered Rush Medical College, but he left without a degree after becoming "infatuated" with botany.[28] His early scientific research, particularly in entomology, was sufficiently impressive that he was named curator of the natural history museum in 1872 and professor of zoology at the Illinois State Normal University in 1875. But only after being appointed chairman of the zoology department at the University of Illinois in 1884 did Forbes receive a somewhat unusual academic degree. "It was also in 1884," Forbes later wrote, "that Indiana University gave me the degree of doctor of philosophy 'on examination and thesis,' entirely the product of private study, as I had taken no academic college course and had no bachelor's degree."[29] Shortly thereafter, at the height of his career, Forbes wrote "The Lake as a Microcosm."

In his famous essay Forbes drew a vivid picture of the aquatic environment and the interacting organisms living there, a picture strikingly reminiscent of Charles Darwin's entangled bank. According to Forbes, the natural order and lack of chaos that characterized this "little community" could be explained by two general ideas.[30] First, there was a *community of interest* even among predator and prey; each prudently acted to maintain the optimal population size of the other. "The interests of both parties," Forbes wrote, "will therefore be best served by an adjustment of their respective rates of multiplication such that the species devoured shall furnish an excess of numbers to supply the wants of the devourer, and that the latter shall confine its appropriations to the excess thus furnished."[31] Second, this well-ordered community had evolved and was maintained by the "beneficent power of natural selection," which, though destructive, promoted the common interests of the constituent species.

Reading Forbes's description of the aquatic community one is struck by the richness of description and the highly literary style of the essay. But one cannot help feeling that Forbes was struggling to develop an appropriate technical language with which to describe the interactions between aquatic plants and animals. In describing these interactions, Forbes relied upon a variety of metaphors: mechanical, organic, political, and economic. Perhaps the dominant image suggested by Forbes's essay is that of nature as a battlefield of each against all. Life in the lake was a "fearful slaughter" of prey by predators, a constant "scramble for food" among competing individuals, and a continual challenge of adapting to an endlessly fluctuating physical environment.[32] Within such an unstable environment few lived to maturity, but Forbes, like Darwin and Spencer, believed that this ceaseless strife was also a mechanism for insuring social harmony and progress:

> In this lake, where competitions are fierce and continuous beyond any parallel in the worst periods of human history, . . . where mercy and charity and sympathy and magnanimity and all the virtues are utterly unknown; where robbery and murder and the deadly tyranny of strength over weakness are the unvarying rule; where what we call wrong-doing is always triumphant, and what we call goodness would be immediately fatal to its possessor,—even here, out of these hard conditions, an order has been evolved which is the best conceivable without a total change in the conditions themselves; an equilibrium has been reached and is steadily maintained that actually accomplishes for all the parties involved the greatest good which the circumstances will at all permit.[33]

This natural equilibrium, however tenuous and imperfect, suggested other images to Forbes, both mechanical and organic. For Forbes, the lake was both a complex machine and an organism.[34] Species within the community were parts of a larger whole, and any change in one had ramifications for other parts and the entire community. From a careful analysis of the stomach contents of black bass, Forbes knew that this important predator relied upon numerous species of insects and crustaceans for food.[35] Directly or indirectly, it depended upon nearly every animal in the lake. This mutual dependence was a general rule in the microcosm, and the intricate interactions among species maintained a regularity and stability in the community as a whole. Life for the individual was a chaotic existence of ceaseless competition and predation, but at the level of the species this led to optimal population size, and at the level of the community it led to a stable equilibrium between predators and prey.

The term *community* that Forbes used to describe the interacting plants and animals in the lake became a fundamental ecological concept during the twentieth century by suggesting a close analogy between human affairs and biological processes. Forbes's lake was not only a microcosm of nature but also a reflection of American society. The economy of nature was dictated by the same law of supply and demand that served as an invisible hand in regulating the marketplace. In both instances, success went to the best adapted and most efficient competitor. "Just as certainly as the thrifty business man who lives within his income will finally dispossess his shiftless competitor who can never pay his debts," Forbes wrote, "the well-adjusted aquatic animal will in time crowd out its poorly-adjusted competitors for food and for the various goods of life."[36] However, Forbes's essay did not evince quite the same optimism in unregulated capitalism as Spencer's early writings. By 1883 Forbes's America had suffered through a recent economic depression, a decade of labor strife, and the uncertainties of a new industrial capitalism increasingly dominated by large corporations. Forbes's aquatic microcosm may have exhibited a harmonious balance, but this balance could be easily disturbed. For example, unpredictable changes in water level might lead to catastrophic death among vulnerable species in the lake.

Forbes's essay, so anthropomorphic in its description of life in a lake, was not aberrant; it was quite typical of proto-ecological literature. For example, a similar style was employed by the botanist Conway MacMillan in his early descriptions of plant communities. MacMillan had studied with Charles Bessey, an eclectic botanist and gifted teacher, who established one of the most influential American schools of ecology at the University of Nebraska.[37] After completing a master's degree at the University of Nebraska, followed by a year of additional study at Harvard and Johns Hopkins, MacMillan was hired as an assistant professor of botany at the University of Minnesota in 1887.[38] Four years later he became chairman of the department and state botanist, posts that he held until he resigned in 1906.

One of MacMillan's duties as state botanist was to complete a botanical survey of Minnesota, part of which appeared as an eight hundred–page list of species, *The Metaspermae of the Minnesota Valley* (1892).[39] Tucked in the middle of this ponderous description of regional flora was a brief discussion of the dynamics of vegetation, a discussion that anticipated important areas of research in early twentieth-century plant ecology. According to MacMillan, the apparent stability and permanence of the plant cover was an illusion; vegetation was actually in a constant state of flux. For economic botanists this fact became obvious when introduced weeds spread quickly at the expense of valuable crops. This was but one example of a general biological and social law: "Every individual plant must make its way in the world. It must either win new territory, maintain what it has already won, or cede its place of abode and growth to some plant better fitted to cope with the conditions peculiar to that particular spot. It thus happens that the flora of any region—that is to say the plant society of the region—is in the same condition of mutual interdependence and mutual competition that we discover in human society."[40] Competition, for MacMillan, was more than simply the war of each against all. It involved a complex set of interactions at a number of levels: the individual, the species, and the plant community as a whole. Individual plants, like humans, competed with one another, but they also cooperated by banding together in mutual self-interest against other groups of plants.

> Each species competes with those around it and in this competion [*sic*] the individuals might be said to stand shoulder to shoulder against the common foe, as may be seen in the united efforts of a human tribe or nation against some warring body. And again groups of species, having perhaps a common line of movement or a common need to be supplied, band themselves together and find arrayed against them other united groups of species competing for the same necessity or striving to move in the opposite direction.[41]

This form of high-level competition was most evident at the boundary between the forest and prairie, where the two great communities—each made up of hundreds of species—engaged in "silent warfare" over contested territory. Thus, like Darwin and Forbes, MacMillan interpreted the struggle for existence broadly. This process occurred at a number of levels and worked hand in hand with cooperation.

Themes and Metaphors

By about 1900 the major themes of ecological discourse were established: change and uniformity, instability and equilibrium, competition and cooperation, integration and individuality.[42] These did not constitute mutually exclusive positions but rather alternative preferences or guiding ideals. If the dominant themes of Darwin's work were competition and change, then this did not necessarily preclude a high degree of uniformity, stability, and interdependence in the natural world. Similarly, the stability of Forbes's aquatic microcosm did not reflect a static equilibrium. For the individual fish, the lake was a hurly-burly of endless strife, and entire populations might be imperiled by unpredictable fluctuations in the environment. Stability was an important concept for Forbes, but he was not a prisoner to some rigid, dogmatic commitment to equilibrium. Like the other proto-ecologists of the late nineteenth century he used this idea flexibly. This is an important point to emphasize. In his book, *Discordant Harmonies,* Daniel Botkin accuses Forbes and other early ecologists of viewing biological rhythms in terms of pendulum-like regularity. Forbes actually used this metaphor in his work, but he was careful to note that the amplitude of the biological pendulum in his aquatic microcosm was constantly altered by disturbing forces.[43] Stability and instability, like other thematic polarities, established a range of possible explanations; they did not define incommensurable positions or dogmatic schools of thought.

The most striking stylistic feature of the literature considered in this chapter is the rich use of metaphorical language. Scientists often dismiss metaphors as mere figures of speech, but it seems likely that these linguistic devices play important, constructive roles in scientific discourse.[44] Metaphors are explanations; when Stephen Forbes described the lake as a battlefield he was providing an explanation of something poorly understood (limnology) in terms of something well-understood by many members of his post–Civil War audience. Metaphorical descriptions suggest analogies, some perhaps strong enough to count as testable predictions. To claim that an aquatic community is an organism suggests a certain level of interdependence among its parts. Forbes, in fact, claimed that *every* member of the community was dependent upon every other.[45] Such extreme interdependence might exist in nature, but it also might not. This claim, a clear reflection of Forbes's organicism, was open to empirical refutation. Metaphors also suggest new questions or lines of research, some of which the originator may not fully recognize. To compare a biological community to a machine in the 1880s would have suggested somewhat different attributes to an audience than the same comparison made a century later during the age of computers. Finally, because they are so suggestive, metaphors may become easy targets for criticism. Opponents can emphasize incongruities and thus potentially discredit not only the metaphor but also the theory it represents. But as explanatory tools, hypothesis generators, heuristic devices, and targets for criticism, metaphors may stimulate the intellectual development of a new area of research.

Ecologists had a number of metaphors to aid in explaining the complex interactions and interdependencies that they encountered in nature: community, organism, and machine. All these were already used in the proto-ecological literature of the late nineteenth century. Despite its obvious limitations—it was inherently anthropomorphic, and it externalized the physical environment—the community became an important concept in ecology. The idea that plants and animals form a kind of community is natural enough. But "community" itself is a changing concept. During the period in which Forbes and MacMillan were writing, America was being transformed from a nation of relatively autonomous, rural "island communities" to an urbanized, industrial culture.[46] The informal social patterns of rural life were giving way to more centralized patterns of authority. Laissez faire, a doctrine that held such appeal for earlier generations, was being replaced with more hierarchical, regulative views of government and economy. The clash of social and economic ideas is reflected in the essays of Forbes and MacMillan. Indeed, the literary style of these writings highlights the influence of social thought upon science. Such influences would be partially, but never completely, obscured by the more technical style of twentieth-century scientific literature.

Today, the idea that plants and animals together form a kind of superorganism is anathema to most ecologists, but it was a popular mode of explanation for early ecologists. The concept of the organism implied organization, stability, and orderly change. As late nineteenth-century writers showed, however, organismal metaphors could also be used to discuss struggle, instability, and random disturbances. The belief that there is a kind of molecular struggle for existence within the body was commonly expressed. In Forbes's aquatic microcosm, which he compared to an organism, the stability of the whole only partially masked the uncertainties of struggle, conflict, and unpredictable change.

That organismal metaphors held such appeal for ecologists should not be too surprising. Organisms are natural objects familiar to all biologists. Through observation, classification, dissection, and experiment on this class of objects the neophyte biologist learns about organic structure and function. Organisms most clearly exhibit those characteristics seemingly unique to life: growth, development, metabolism, and reproduction. For ecologists during the early twentieth century organismal metaphors may have been particularly compelling for two other reasons.[47] Physiology, the queen of the life sciences during the late nineteenth century, had established an enviable repertoire of exact experimental techniques for studying organisms. Early ecologists often looked to physiology as a model of rigorous, experimental science. The cell theory, well-established by the late nineteenth century, provided a conceptual framework for discussing organic relationships, both in terms of part/whole and structure/function. Was it not reasonable to suppose that similar relationships existed between the individual organism and the biological community?

Mechanical metaphors played a similar, if somewhat less important, role in early ecology. In fact, the ideas of nature as a superorganism and nature as a machine were often used interchangeably.[48] The apparent equilibrium of a biological community could be compared to either the self-regulation in an organism or a machine controlled by a governor. In his essay on the aquatic microcosms, Forbes implied this relationship between self-governing machines, organisms, and biological communities. However, he did not fully develop his mechanistic descriptions of the lake. For much of the early history of ecology, mechanical metaphors remained an intellectual undercurrent. Only with the development of complex cybernetic systems during World War II did metaphors in ecology shift from organic to mechanical.[49]

Community, organism, machine: all these metaphors are consistent with Darwin's idea that nature forms a web of complex relations. But one might interpret Darwin in another way: perhaps his entangled bank is only an illusion. Perhaps the apparent order of nature is owing to populations independently adapting to a common physical environment. A community, if you care to use such a term, may be little more than a collection of autonomous populations that just happen to occupy the same place at the same time.[50] This "individualistic" view of nature, absent from the proto-ecological literature of the late nineteenth century, remained a minor eddy in the mainstream of early ecological thought. It too had to await the Second World War before it gained large numbers of adherents.

2

A Rational Field Physiology

There can be little question in regard to the essential identity of physiology and ecology.

—FREDERIC EDWARD CLEMENTS, *Research Methods in Ecology*

THE IDEA OF A GROUP of interdependent organisms, what Stephen Forbes referred to as a "community," became a central concept in ecology. Although Forbes and others discussed it during the late nineteenth century, the biological concept of community was most fully developed by midwestern botanists beginning about 1900. These botanists, among the first scientists to self-consciously identify themselves as ecologists, often claimed to be revolting against the genteel tradition of collecting and identifying specimens. Traditional botany, the young Frederic Clements claimed, "lends itself with insidious ease to chance journeys or to vacation trips, the fruits of which are found in vague descriptive articles."[1] In contrast, ecology was to be a rigorous, experimental science dealing with biological processes and their causes. Indeed, for Clements ecology was "nothing but a rational field physiology."[2] Midwestern plant ecologists were never quite so revolutionary as they sometimes claimed, and others were not always impressed with their vision of a new experimental science.

Critics sometimes dismissed plant ecology as a glorified agricultural science, but Clements and his contemporaries were interested in more than applied botany.[3] Ecologists, at the turn of the century, also were passionately interested in the broader biological problems of adaptation, development, and distribution. For these ecologists change was the primary characteristic of the natural world, and they called for a dynamic, process-oriented science that could explain this change. The physiological perspective that these biologists embraced had a pervasive influence upon the later development of ecological thought. Not only did this perspective suggest innovative methods for studying nature, but it also provided an explanatory framework that was both organic and mechanistic.

At the turn of the century, the midwestern United States provided a fertile environment—both physically and intellectually—for ecological studies. Most ecologists believed that fundamental biological problems could be studied best in natural laboratories: forests, lakes, dunes, and prairies. The frontier may have been coming to a close in 1900, but ecologists could still find pristine environments to study, and these natural laboratories were readily accessible from midwestern universities. Intellectually, the Midwest also provided a stimulating environment for the young science of ecology. New institutions such as the University of Chicago were breaking with traditional approaches to education.[4] As universities in the Midwest proliferated, biologists with new ideas about the nature of their science had unique opportunities to shape departments and research programs along nontraditional lines.[5] In a more subtle way, the midwestern environment may have influenced the way that ecologists approached their work. Writing at the turn of the century, the historian Frederick Jackson Turner suggested that the frontier was a powerful force capable of shaping human character.[6] Using a variation of this frontier thesis, Paul Sears later argued that ecological theory was significantly influenced by the midwestern environments within which early ecologists worked.[7] One need not accept this environmental determinism too literally to imagine the powerful impress that dunes and prairies made on the minds of early ecologists. It is perhaps not too surprising that the themes informing the historical writings of Turner at the University of Wisconsin—organic development, adaptation, evolution, and the interaction of organisms with their physical environments—also appeared prominently in the early literature of plant ecology.[8]

Henry Chandler Cowles and the Life History of Sand Dunes

Typical of this new breed of midwestern botanist was Henry Chandler Cowles. After graduating from Oberlin College in 1893, Cowles began graduate studies at the newly opened University of Chicago, an institution where he remained throughout his professional career.[9] During its early years, the University of Chicago was an exciting intellectual environment for an aspiring young scientist. President William Rainey Harper was luring many outstanding scientists to develop new research and teaching programs at the university.[10] Cowles initially began studying geology under T. C. Chamberlain, a noted geologist recently recruited from the University of Wisconsin. However, captivated by professor John Merle Coulter's discussions of plant life on the sand dunes of Lake Michigan, Cowles soon switched to botany. His dissertation described the long-term process of vegetational change, or succession, as it occurred on the dunes.[11]

Cowles's research was a stimulating mix of careful observation and rather speculative theorizing. He described in great detail the environmental conditions and the various "plant societies" that existed in the dunes: perennial herbs, shrubs, heath, coniferous forest, and deciduous forest. He then arranged these societies into developmental series. As Cowles envisioned the process, small embryonic dunes formed on the beach as sand washed up on shore. Some of these dunes were stabilized by colonizing plants, whose fibrous roots trapped the sand and prevented it from blowing away. The dune and its community of plants formed a symbiotic relationship, and over several hundred years both developed in a fairly predictable manner. Given proper environmental conditions the terminus of this developmental process, the climax community, was a large sandy hill covered with a deciduous forest dominated by beech and maple trees.

Cowles's research was inspired by earlier investigations of oceanic dunes in northern Europe, particularly those conducted by the Danish botanist Eugenius Warming.[12] Cowles shared Warming's enthusiasm for a physiological approach to the study of plant communities. Both men were committed to explaining the structure and distribution of plant communities in terms of the relationship between physical factors in the environment and physiological adaptations in plants. However, in a number of ways, Cowles's study differed from its European model. Despite his new-found love for botany and his interest in physiological adaptation, geology continued to hold a powerful attraction for Cowles. It provided a model for creating the new, dynamic ecology typified by his study of succession: "Such a study is to structural botany what dynamical geology is to structural geology. Just as modern geologists interpret the structure of the rocks by seeking to find how and under what conditions similar rocks are formed today, so ecologists seek to study those plant structures which are changing at the present time, and thus to throw light on the origin of plant structures themselves."[13] Geology also provided an important part of Cowles's explanation of succession. Changes in the topography of a region were the ultimate forces causing vegetational changes. Cowles's perspective was more than geological, however; he also employed the organic metaphors so common in American intellectual life at the turn of the century.

Cowles never claimed that a plant community was an organism, but organic analogies were common in his writing. Throughout his classic paper Cowles portrayed ecological succession as a developmental process. The dune began as an embryo, passed through a series of developmental stages, attained maturity, and eventually died. The fact that this complicated process did not always occur in exactly the same way did not alter the fact that the ecologist, studying a number of dunes, could describe the idealized life history of a dune.

Idealization is an important step in theory construction, and Cowles moved from observation to abstraction in a sophisticated manner. No single dune necessarily went through a particular series of developmental stages; the physical forces controlling succession were too unpredictable for that. "The simple life-history just outlined is the exception," Cowles wrote, "not the rule. . . . These processes of deposition and removal, dune formation and dune destruction, are constantly going on with seeming lawlessness."[14] Nonetheless, the simplified developmental scheme provided an explanatory framework for understanding the relationships among the various individual dunes making up a "dune complex." Using the metaphor of ontogeny Cowles systematized the seemingly chaotic changes in the soil and vegetation of the Indiana sand dunes.

Cowles's sand dune, like Darwin's entangled bank and Forbes's aquatic microcosm, was a scene of seemingly chaotic, lawless change, and at the same time, the site of orderly, law-governed processes comparable to those in a developing organism.[15] Both the predictability and the unpredictability of succession could be explained by the causal web underlying this developmental process. What Cowles described as a symbiotic relationship between the dune and its community of plants was not a simple interaction, but rather a shifting balance between two powerful agents of change. Plants could capture, stabilize, and modify the dune, but this outcome was not inevitable. Even though the dune provided the necessary resources for the development of vegetation, it also constituted a harsh and unpredictable environment for plants. Intense sunlight reflecting off the sand, lack of moisture, and a nutrient-poor soil, all provided extreme challenges to life on the beach. Most important, however, was the destructive sand-blasting effect of winds blowing off the lake. Living tissue could hardly withstand such abrasion. The windward branches of trees were often stripped bare of soft tissue, leaving only a network of tougher fibers. Even more dramatic was the destructive effect on nonwoody plants. "Fleshy fungi have been found growing on the windward side of logs and stumps completely petrified, as it were, by sand-blast action"; Cowles noted, "Sand grains are imbedded in the soft plant body and as it grows the imbedding is continued, so that finally the structure appears like a mass of sand cemented firmly together by the fungus."[16] Wind not only destroyed individual plants, but it could also destroy whole plant communities. Under the influence of wind the sand dune was never a completely stabilized environment. Rather, this restless maze could break away, uprooting an established plant community and burying other communities as it advanced across the beach. This was not an uncommon event; the area was dotted with the "graveyards" of forests buried by wandering sand dunes. At any given time, therefore, the relationship between dune and plant community could develop into one of several possibilities. An uninhabited, wandering dune might be captured and colonized by plants. Together the dune and its inhabitants might grow and develop toward a climax community. Or the dune might break away from the stable relationship and destroy its symbiotic living partner in the process.

Competition played an important role in succession. The regularity of succession could be largely explained by replacing established species with better adapted competitors. For example, in certain moderately moist areas pines were replaced by oaks, not because the pines were poorly adapted to moist soil, but because oaks were better adapted.[17] Unlike Forbes's lake, however, competition on the sand dune was less a matter of struggle among individuals, than a struggle between the individual and its physical environment. The extreme conditions found on the dune posed a continual challenge to plant life, and only a few, well-adapted forms could meet this challenge. To successfully capture a wandering dune, a plant had to have an extensive system of roots to trap and hold sand. But even for those plants with networks of fibrous roots there was an ongoing struggle to hold sand against the eroding force of the wind. To be successful, therefore, the plant had to be capable of growth even when buried by the shifting sand. And, most important, it had to survive periodic exposure of its roots as the dune was eroded by wind. "In short," Cowles concluded, "a successful dune-former must be able at any moment to adapt its stem to a root environment or its root to a stem environment."[18] Thus, on the sand dune, adaptation often meant adaptability to a relentless and unpredictable physical environment. Cowles was particularly interested in the physiological adaptations that allowed individuals to successfully compete with this physical environment, but he was also interested in cooperation among individuals within a society of plants. A single plant was generally no match for a moving sand dune; the successful capture of the dune required the cooperative effort of many individuals.[19] Competition and cooperation, important causes of succession, both occurred on the dune.

Cowles's study of succession on the sand dunes of Lake Michigan became a classic in the literature of ecology. Its careful description and analysis became a model for Cowles's students and a source of inspiration for later ecologists, particularly those interested in ecosystems. Half a century later, Jerry Olson, another University of Chicago graduate student, returned to study succession on the dunes. His award-winning research placed dune succession within an explicit ecosystem context.[20] At about the same time, Eugene Odum, the dean of ecosystem ecologists, favorably compared Cowles's influence in ecology to that of Gregor Mendel in genetics.[21] Like many of his fellow ecosystem ecologists, Odum considered succession a fundamental ecological process.

Odum emphasized the dynamic nature of ecosystems, and his discussions of succession were rooted in the same physiological perspective that so attracted Cowles. Despite his enthusiasm for physiology, Cowles himself never fully developed this approach to research. His early work was primarily descriptive, and later in his career he published relatively little original research of any kind. As a young man, Cowles suggested that a plant community was analogous to an organism and that its internal processes could be studied physiologically, but he never made the transition to a truly physiological ecology. Ecology, for Cowles, remained firmly within the domain of natural history. Hiking across the dunes toward Lake Michigan, the ecologist could walk backwards in time, retracing the developmental history of the plant community.

Frederic Clements: The Physiological Perspective in Ecology

Intellectually, Cowles and Clements had much in common, and they both played important roles in establishing plant ecology in America. However, it is difficult to imagine two individuals so profoundly different in personality and scientific style. Cowles was a popular teacher whose warmth of personality and sense of humor were legendary.[22]

He attracted a large and devoted group of students who continued the research program that he began during the 1890s. To a great extent, Cowles's legacy rests upon the intellectual lineage that he started at the University of Chicago. In contrast, Clements was often arrogant, priggish, and distant, inspiring little warmth even in those who knew him best. Although he trained a few students, most of his career was spent outside academia as a research associate at the Carnegie Institution of Washington. Unlike Cowles, who taught much but wrote little, Clements's influence arose from his voluminous writings; his books and articles touched on virtually every topic in ecology. Something of these differences is captured in photographs of the two great ecologists: Cowles, always looking a bit rumpled, often with a battered hat on his head and a boyish grin on his face, is a study in contrast with the stiff, neatly pressed, and unsmiling Clements (figures 1 and 2). Clothes may not make the man, but these contrasting portraits mirrored fundamentally different intellectual styles. Cowles often referred to the chaotic state of ecological thought as it existed early in the twentieth century, a situation he seemed to accept as inevitable. Ecology was to be a search for natural laws, but the nature that Cowles encountered on the sand dunes of Lake Michigan often appeared capricious.[23] In Cowles's ecology there remained a tension between order and disorder. Clements abhorred chaos in ecological thought as much as in his puritanical personal life. Out of his search for order emerged a theoretical ecology that was sweeping in its breadth and audacious in its simplicity.

When Clements entered the University of Nebraska in 1891, the botany department was gaining a national reputation. Charles Bessey was attracting bright young students and training them in the "new botany." Loosely patterned on a German educational model, Bessey's new botany emphasized experimentation and laboratory techniques, particularly the use of the microscope. However, according to Ronald Tobey, there was something uniquely American in Bessey's pragmatic approach to biological education.[24] Like other American intellectuals during the late nineteenth century, Bessey and his students were in revolt against what they perceived as the sterility of traditional education. Botany was to be learned not through rote memorization of textbooks but through experience in the laboratory and in the field. And it always had an eye toward the practical problems of agriculture and forestry.[25]

Bessey's new botany with its emphasis on experimentation and laboratory technique shaped Clements's approach to the study of ecology. Throughout his career, Clements liked to portray himself as a radical educator and a scientific innovator.[26] However, his vision of an experimental, physiologically oriented ecology did not crystallize immediately. Like Cowles's study of sand dunes, Clements's early research was descriptive. His doctoral research, done jointly with fellow student Roscoe Pound, was a fairly conventional study of regional vegetation. Inspired by the earlier geographical studies of Oscar Drude, Pound's and Clements's *The Phytogeography of Nebraska* catalogued species, described plant formations, and correlated these formations with general features in the environment.[27] But unlike the later Clements, he made no consistent attempt to measure environmental factors or to investigate causal relationships through experimentation. The book was a transitional work; the authors used ecology as one of several useful perspectives from which to study plant distribution. By 1905, Clements, then an associate professor at the University of Nebraska, reversed this relationship. In his first major ecological work, *Research Methods in Ecology,* a book that brought him international recognition, plant geography was presented as a small part of a more inclusive and rigorously experimental science of ecology. In the jargon of his new book, the descriptive geographical research that he had done as a graduate student was *reconnaissance,* a necessary but rather mundane prelude to more ambitious ecological experimentation.

The Plant Community as Organism

Running through Cowles's classic study of sand dunes is the idea that the plant community is like an organism; however, he never fully developed this suggestive analogy. Frederic Clements made this idea explicit and used it as a central concept in his theoretical ecology. For Clements the plant community really was a "complex organism," and succession was its life cycle. In turning to these organismal ideas, Clements avoided the naive anthropomorphism that Herbert Spencer succumbed to in the "Social Organism" and the traces of romantic imagery that lingered in the writings of Stephen Forbes, Conway MacMillan, and Cowles. Clements saw himself as a tough-minded professional struggling to create a technical vocabulary for ecology. The "complex organism" was not just a suggestive image; it was an important theoretical term for ecologists.

What did Clements mean when he claimed that the community is a kind of organism? It most certainly was not an organism in the same sense as a vertebrate animal or even a higher plant. What Clements seemed to have in mind as models for the community-organism were much simpler plants and animals, perhaps what we would refer to today as protists.[28] But even these models were not to be taken too literally. The similarities between simple organisms and complex organisms were not to be found in naive isomorphisms; parts of a forest were not precisely comparable to any anatomical structure. The simple organism and the community had in common a number of general biological characteristics. The community was an organic entity made up of interacting parts, much the way an individual was composed of interacting cells. It had spatiotemporal continuity, and it developed in a fairly predictable manner. Finally, the community had a kind of physiology, a set of processes through which it interacted with the physical environment to maintain a dynamic equilibrium. The community was capable of adapting just as any organism did.

Given the historical context within which he was working, Clements's suggestion that a plant community is a kind of organism was quite unremarkable. Organismal analogies had long been popular, and they continued to be characteristic of the intellectual landscape of early twentieth-century America.[29] Clements, however, was unusual in the way that he tied his organismal concept to a broader physiological perspective. Above all, Clementsian ecology was the study of *processes*, and physiology provided a successful model for ecological methods and explanations. Clements considered physiology to be the paradigmatic example of a rigorous, experimental science. If ecology was to become "a rational field physiology," then ecologists would need to develop equally rigorous methods for studying plants outside the laboratory. One purpose of *Research Methods in Ecology* was to acquaint ecologists with new quantitative and experimental techniques. Physiological theories, notably cell theory, provided an explanatory model for ecologists. Just as the physiologist could explain the functioning organism in terms of cellular activity, Clements hoped to explain the functioning of the "complex organism" in terms of the activities of its parts.

Late in his career, Clements dabbled in philosophical holism, but his physiological perspective actually reflected an extreme form of mechanistic reductionism. At all levels—individual, species, or community—Clements explained change in terms of simple, stimulus-response reactions. The physical environment acted and organisms reacted. By the time that Clements began writing, physiologists were moving away from such simplistic explanations, and the physiologists who read *Research Methods in Ecology* were universally hostile toward it.[30] Certainly today it is easy to smile at his naive mechanistic ideas.

Although he was wrong in the details, Clements provided future ecologists with a compelling intellectual approach to research. As we see in later chapters, other ecologists also looked to physiology, both for methodological and explanatory models. For now, however, we must examine Clements's physiological approach in greater detail.

Adaptation, Evolution, and Succession

The operation of Clements's organismal concepts and his broad physiological perspective can be seen in the way that he explained individual adaptation, speciation, and succession—three processes that he considered closely related. In *Research Methods in Ecology* he discussed several examples of adaptation to the physical environment. For example, he described what he believed to be the causal chain linking light intensity, photosynthesis, and the gross morphology of leaves.[31] An increase in light intensity (a "stimulus") caused a proportionate increase in the rate of photosynthesis (a "response"). Within the cell, this caused an increase in the number and size of starch grains. More important, the number of chloroplasts increased, optimizing the absorption of light.[32] These intracellular changes caused changes in the arrangement of cells and tissues and ultimately led to gross morphological changes in the leaf. In retrospect, it is easy to dismiss this as speculation, and Clements himself admitted that no conclusive experimental evidence supported his hypothesis. However, he did cite some indirect evidence to support his claims. It is a well-known fact that shaded leaves and sun-exposed leaves, even on the same plant, frequently exhibit distinct morphological features. As Clements attempted to demonstrate through microscopic examination, these gross changes were correlated with changes at the cellular and subcellular levels.

From this example it is clear that adaptation meant something more than the Darwinian fit between organism and environment. For Clements it also meant the physiological process of adjustment of which all organisms are more or less capable. This physiological adaptation had important evolutionary implications. As a neo-Lamarckian, Clements believed that environmental changes could directly cause the evolution of new species. He attempted to demonstrate the inheritance of acquired traits by transplanting low-altitude species into experimental gardens located on Pike's Peak in Colorado. As the transplants developed in their new habitat they took on characteristics typical of alpine plants, and in some cases Clements claimed that they became indistinguishable from species native to the mountain.[33] He concluded that by modifying the environment of a plant he could artificially induce speciation within several generations.

Clements's experimental neo-Lamarckism was within the mainstream of evolutionary biology during the first decade of the twentieth-century when he began his work, but much less so in 1945 when he died. His later career, discussed in chapter 3, increasingly became a quixotic attempt to document his evolutionary claims. The point I stress here, however, is not the long-term significance of Clements's evolutionary views, but rather the breadth of his physiological perspective. Beginning early in his career Clements proposed a unified mechanical scheme to explain both physiological and evolutionary adaptation. During the course of a single generation, individual plants adapted physiologically to changes in the environment. Over the course of several generations, species evolved in response to persistent changes in the environment. According to Clements, the same type of reasoning could explain the successional changes in plant communities.

Clements's reputation rests primarily upon his contribution to the study of succession. He outlined his theory of succession in *Research Methods in Ecology* (1905) and expanded these ideas in his most important book, *Plant Succession* (1916). This massive tome immediately became the definitive work on the subject, and today it remains a point of departure for many discussions of succession. For Clements, succession, a complex process of development, led from an embryonic community, through a series of stages, to the mature climax community. Despite its complexity, this developmental process could be reduced to a few simpler processes: plants invaded an area, they competed, they reacted to the physical environment, and they modified it.[34] Each process could be understood in terms of simple stimulus-response mechanisms.

Succession began when species invaded a previously uninhabited area. The success of the various migrants in establishing themselves depended upon their competitive abilities. But this was primarily indirect competition, more physical than biological. "Competition," Clements wrote, "is purely a physical process. With few exceptions . . . an actual struggle between competing plants never occurs. Competition arises from the reaction of one plant upon the physical factors about it and the effect of these modified factors upon its competitors."[35] Thus, for example, water absorbed by one plant was unavailable to others. Although such competition acted only indirectly on individuals, it played a decisive role in structuring the community: "The inevitable result is that the successful individual prospers more and more, while the less successful one loses ground in the same degree. As a consequence, the latter disappears entirely, or it is handicapped to such an extent that it fails to produce seeds, or these are reduced in number or vitality."[36]

The composition of a community at a given time reflected the relative adaptation of various species to a particular physical environment. However, this composition was not static; each stage in succession was characterized by a different set of species. Invasion and competition alone could not explain this dynamic nature of the community; another process was also involved. As plants reacted to their physical surroundings they modified important environmental factors. For example, by shading previously bare ground, pioneer species increased the moisture of the soil. This change in an important physical factor allowed new invaders to become established, which then altered the competitive balance in the community. Thus species were often replaced as an indirect result of the very environmental changes that they had caused. "The reactions of the pioneer stages may be unfavorable to the pioneers," Clements wrote, "or they may merely produce conditions favorable for new invaders which succeed gradually in the course of competition, or become dominant and produce a new reaction unfavorable to the pioneers."[37] In either case, the pioneer species were replaced by a new assemblage of plants. This developmental process continued until a climax community was established, a community that he described as "more or less permanent." Unless some external disturbance disrupted this process, the eventual establishment of the climax was as inevitable as the development of an adult plant from a seed.[38]

The persistence of the climax community could be explained in a number of ways. Through successive modification of physical factors such as light and moisture, the community progressively stabilized its environment. In contrast to early successional stages, which were in a constant state of flux, the climax was in equilibrium with its physical environment. As a result, the climax community formed a kind of biological barrier to further invasion; potential invaders could rarely compete successfully with established species. Once formed, this climax could persist indefinitely. According to Clements, "such a climax is permanent because of its entire harmony with a stable habitat. It will persist just as long as the climate remains unchanged, always providing that migration does not bring in a new dominant from another region."[39]

Equilibrium and the Climatic Climax

No aspect of Clementsian ecology has proven so controversial as his ideas on climax. A later generation of ecologists reacted against what they referred to as the Clementsian "monoclimax" concept, the idea that within a given climatic region succession always ends in a single type of community. Historians have also criticized Clements for his apparent determinism. For example, J. Ronald Engel argues that Clements accepted Herbert Spencer's deterministic worldview, a view that Engel compares unfavorably with Henry Chandler Cowles's belief in the flexibility and indeterminacy of succession.[40] Clements was a determinist of sorts, but the monoclimax is a parody of Clementsian thought. Of course, a parody requires something to imitate, and Clements's pedantic style lent itself to easy ridicule. But critics have ignored the important qualifications that Clements added to his theory. Clements was not some naive, armchair theoretician. Although deeply committed to his theories of succession and climax, he was a keen observer who knew about the complexities of nature.

Henry Chandler Cowles once characterized the developing equilibrium between plant community and physical environment as "a variable approaching a variable, rather than a variable approaching a constant."[41] Clements refused to go that far, but he, too, held a dynamic concept of equilibrium. The community was an organism, and like all organisms it was constantly adjusting to environmental fluctuations. "The most stable association is never in complete equilibrium. . . . Even where the final community seems most homogeneous and its factors uniform," Clements warned, "quantitative study by quadrat and instrument reveals a swing of population and a variation in the controlling factors. Invisible as these are to the ordinary observer, they are often very considerable."[42] External factors might drastically upset the equilibrium between the community and its environment. No community, even the climax, was completely closed. There was always the possibility that a foreign species might successfully invade the community.[43] Natural disasters or human interference could also modify climax patterns. Forest fires, logging, erosion—any one of these might damage or destroy the climax. In these damaged areas succession would begin again, but the overall result would be a landscape resembling a mosaic of climax and subclimax vegetation.

Clements's little-known study of forest fires in Estes Park is a case in point.[44] Periodic fires in the area often prevented the establishment of the theoretical climax forest, a forest dominated by Engelmann

Spruce. Lodgepole pine, an early successional species particularly well-adapted to seeding burned areas, sometimes persisted as the dominant species indefinitely. This was also true of aspen, which could quickly regenerate from underground roots after a fire. From experiences such as this, Clements knew that nature was complex; vegetation formed a mosaic, not a monotonous continuum. But ecology had to do more than simply describe this mosaic; it had to explain it. For Clements the concept of organic development provided the explanation.

The Individualistic Challenge

One could, of course, argue that Clements's particular explanation was badly misleading, that, even ignoring complicating factors, succession is not analogous to development and the community is not analogous to an organism. Precisely this charge was made by Henry Allan Gleason.[45] Like Clements, Gleason was born and raised in the Midwest. After completing bachelor's and master's degrees at the University of Illinois, he went on to receive the Ph.D. from Columbia University in 1906. He then taught at the University of Michigan for ten years, after which he moved to the New York Botanical Garden, where he remained for the rest of his career. Until the late 1960s he continued to publish papers on diverse topics in ecology, taxonomy, and plant geography.

In a series of short papers written over a twenty-year period, Gleason put forward what he referred to as the "individualistic concept" of the plant community.[46] Gleason claimed that the similarities between succession and ontogeny were superficial; succession was not a developmental process in any meaningful sense. Rather, this much less deterministic process depended to a large extent upon random events. As a consequence, the plants found in a particular area did not form an organic entity, but simply an assemblage of individuals.

Gleason based his reasoning on three premises. Environmental factors, particularly physical ones, always vary both in space and time. Each species of plant has a range of environmental tolerances, as does each individual member of the species. Plants tend to disperse seeds randomly. Thus, according to Gleason, the distribution of plants in any particular area was the result of fortuitous immigration and environmental selection. As the environment changed, so did the distribution of various species of plants. Gleason suggested, therefore, that succession was nothing more than a statistical replacement process.

Better adapted species gradually replaced less well adapted ones, but replacement also occurred by the more random process of seed dispersal.

Given this indeterminism in plant distribution, what then was the status of the plant community? Certainly not objective units, Gleason argued, but "merely abstract extrapolations of the ecologist's mind".[47] If environmental factors varied continuously in space and time, if every species (indeed, every individual) had a unique range of environmental tolerances, and if immigration were fortuitous, then ecologists would often find plant communities with poorly defined boundaries. In general, species would be distributed independently across the landscape, and any two geographical areas, no matter how small, would contain slightly different assemblages of species. For Gleason, the biological community was little more than a coincidental assemblage of independent species sharing an area arbitrarily defined by the ecologist. "In conclusion," he wrote, "it may be said that every species of plant is a law unto itself, the distribution of which in space depends upon its individual peculiarities of migration and environmental requirements."[48]

Much has been made of the "Clements-Gleason controversy." Gleason liked to portray himself as an "ecological outlaw," and a later generation of ecologists popularized this image of Gleason as the embattled critic of ecological dogma, a critic whose ideas were later vindicated by rigorous experimental testing.[49] Such stories should not be given too much credence; within the historical context of pre–World War II ecology, the controversy amounted to very little. In his two most famous articles, Gleason never mentioned Clements's work directly. And Clements, who held Gleason in rather low regard, never responded to the younger ecologist's critiques.[50] But why was the individualistic concept so unpopular prior to World War II? To claim that ecologists of the 1920s and 1930s were dogmatically committed to Clementsian ecology is historically false. But even if true, it would not serve as an intellectually satisfying historical explanation. The questions still remain: Why were ecologists so committed to Clementsian organicism? And why did they find the individualistic concept unsatisfactory?

I suggest a number of alternative explanations. One explanation is institutional. Other ecologists supported the individualistic concept,[51] but prior to World War II none of these biologists developed an effective research program. Gleason may have thought that the individualistic concept was important, but he did not pursue it very far. He never collected data to support his claims, and his theoretical writings on the subject were limited to three short papers. As we shall see in chapter 3, Clements was a more astute "empire-builder" than his critics. During the 1920s he mustered the considerable resources of the Carnegie Institution of Washington to effectively promote his organismal ideas.

Politics was only part of the equation, however. There were also sound intellectual justifications for rejecting Gleason's individualistic concept. Lack of data was one. To an empiricist, Gleason's theoretical sketches would have compared rather unfavorably with the massive body of information collected in Clements's *Plant Succession.* Theoretically too, Gleason's argument was flawed. The suggestion that communities are not organisms *because* they lack distinct boundaries is an obvious non sequitur. Humans and some other animals may have rather definite external boundaries, but many other types of organisms do not; this distinction was pointed out by organismal thinkers both before and after Gleason's day.[52] In a broader sense, Gleason's concept lacked a convincing theoretical justification. For many biologists today, the individualistic concept is attractive because it fits nicely with recent theoretical trends in population ecology. Gleason was no population ecologist, however, and the apparent modernity of his ideas is deceiving. Evolution may have been implicit in his arguments, but Gleason did not use natural selection to justify his claim that ecology could be reduced to the activities of independent individuals. Nor did he refer to the theoretical population ecology or population genetics that was beginning to develop during the period in which he was writing. Clementsian ecology, whatever its problems, did have a well-developed theoretical foundation. In short, without substantial data or a satisfactory theoretical foundation, Gleason was asking his readers to abandon an apparently successful approach to research. Not surprisingly, even some ecologists who were highly critical of certain aspects of Clementsian ecology refused to embrace the individualistic concept.

One can get some sense of the response to Gleason's ideas from a student notebook of Raymond Lindeman.[53] Lindeman went on to write one of the formative papers in ecosystem ecology, but in 1937 he was a beginning graduate student in W. S. Cooper's plant ecology course at the University of Minnesota. Cooper's course was a mixture of lecture and discussion; the discussions often continued informally at the Cooper home. Cooper, a University of Chicago graduate, stressed the importance of carefully describing successional patterns and discovering the laws governing these patterns. The organismal concept was important, too, although Cooper believed that Clements pushed it too far. Gleason's paper was thought provoking but flawed. From the class discussion, Lindeman wrote down a long list of criticisms. Gleason's emphasis on accident and coincidence seemed to rule out the possibility of general laws of succession. The idea that a community is only a chance collection of individuals seemed unreasonable. Indeed, Gleason's heavy emphasis on randomness and indeterminacy seemed unjustifiable. Most important, the idea that indefinite boundaries vitiated the concept of community seemed untenable. "Do not transitions occur between everything?" Lindeman noted. "Then why throw out the idea of community because it can't be sharply defined?" Like most of his contemporaries Lindeman did not accept Clementsian ecology in its entirety, but he did believe that useful parallels existed between organisms and communities. From Cooper, he imbibed modified Clementsian ideas of succession, climax, and equilibrium, and he accepted the Clementsian notion that ecology is the study of physiological processes—ideas that later found expression when he attempted to define the scope of ecosystem ecology.

Succession and the Physiological Perspective in Ecology

The history of scientific controversies is not always neat and tidy, with clear winners and losers. The continued influence of Clementsian ideas, even after their apparent defeat, is a case in point. Shortly after World War II, Gleason's individualistic concept was partly vindicated by the field studies of a number of ecologists. Using a new technique, gradient analysis, John T. Curtis and his students at the University of Wisconsin and Robert H. Whittaker at the University of Illinois demonstrated that in many cases communities lacked clearly defined boundaries. As Gleason had predicted, populations scattered along environmental gradients formed continua rather than discrete units. Succession did not appear to follow neat linear sequences, and the climax seemed an indefinite mixture of species. "Climaxes are relative," Whittaker wrote, "and there are all degrees of climaxness."[54] For Whittaker and most later ecologists, the climax was a mosaic of vegetation, an entity definable only in statistical terms. During the 1950s Whittaker became an outspoken critic of Clementsian ecology. He and others often portrayed their work as destroying a Clementsian paradigm.[55] But Whittaker's work itself provides a good example of the tenacity of Clementsian ideas. When he discussed the structure of communities, Whittaker was an avowed Gleasonian individualist.

However, when he discussed the *processes* that occur in communities and ecosystems, Whittaker often slipped back into more organic or physiological descriptions: populations were parts of a larger whole, and each played a specific functional role to maintain the integrity of that whole.[56]

Much of Clementsian ecology has not stood the test of time. His continued belief in the inheritance of acquired traits, long after it was rejected by most other biologists, was aberrant. His mechanistic notions of cause and effect were considered simplistic even by many of his contemporaries. His insistence that succession is always progressive was also rejected by many ecologists of his day. His ideas of climax and the organic unity of the community were more influential, but they too have been modified or abandoned. Yet, despite all this, Clementsian thought has been enormously influential. As even his critics admitted, the very scope and systematic nature of Clements's work integrated ecological thought, and it stimulated both further research and criticism.[57] More important, Clements emphasized the importance of process in ecology, and he suggested a useful physiological perspective for studying it. This had a powerful influence on the development of ecology.

Few ecologists after World War II believed that a community or ecosystem really was an organism, but in important ways they continued to believe that these higher level systems behaved somewhat like organisms. Succession was the paradigmatic example. Although the Clementsian explanation was wrong in its details, the general idea that succession is a developmental process continued to serve as an important heuristic argument and a useful framework for explanation.[58] The physiological perspective suggested other important analogies between organisms and higher level systems. After World War II ecosystem "metabolism" and "homeostasis" became important areas of ecological research. Clements never considered these ideas, but they fit neatly into his general view that ecology was to be "a rational field physiology."

3

An Ambiguous Legacy

One's success as a scientist can be measured more by the number of people he or she puts to work on new problems than by the correctness of specific research results.

—David M. Raup, *The Nemesis Affair*

The man who states a general theory which leads subsequent workers along the most fruitful lines of research performs a service which is fundamental to the progress of science.

—A. G. Tansley, "Frederic Edward Clements, 1874–1945"

Frederic Clements is an enigmatic historical figure. Universally recognized as one of the founding fathers of ecology, he has, nonetheless, become a convenient "fall guy" for some modern ecologists.[1] During his lifetime Clements's opponents ridiculed his ideas by characterizing them as "flights of fancy," "fairy tales," and "laughable absurdities."[2] Yet much to the consternation of modern critics, these same Clementsian ideas persist in modified form today.[3] How can one explain this ambiguous legacy? The story of the Clements-Gleason controversy, so popular among ecologists today, provides few insights. Indeed, the answer to this question is not found in intellectual comparisons removed from social context. The ambiguities surrounding Clements's historical reputation are better explained by considering the fate of the research group that he formed during the second decade of the twentieth century.

In its details, Clementsian ecology was badly flawed. But being wrong, perhaps even being egregiously wrong, is not antithetical to good science. Most scientists are wrong most of the time, and even great scientists turn out to be wrong much of the time. What is really important in science is identifying an important problem or proposing a fruitful set of ideas that then become the roots for a new line of research. Often, as this new intellectual lineage develops, the original ideas are refined, modified, or replaced. Historians, sociologists, and philosophers have emphasized the important role that small groups of interacting scientists (i.e., research schools or research groups) play in this developmental process.[4] The research group provides an established scientist with an array of resources that greatly expands the scope of scientific work. It also provides the scientist with a potential means for promoting the development of original ideas. To the extent that a scientist can successfully recruit bright young workers, committed to a particular line of research, he or she can establish an intellectual tradition. Failing this, a scientist's ideas are at the mercy of the broader scientific community, where they are frequently ignored. Even if they prove influential, without an active research group to nurture the growth of nascent ideas the scientist has little control over how other independent scientists will interpret and modify these ideas.

Clements's association with the Carnegie Institution of Washington dramatically illustrates both the potential resources provided by the research group and the problems of fully exploiting these resources. His career demonstrates the way that personality can interact with intellectual and political factors in determining the fate of a line of research. The Carnegie provided Clements with great potential resources, both financial and institutional. He skillfully used these to promote his ecological ideas, and these ideas influenced many ecologists. But Clements was less successful in establishing an effective, enduring research group at the Carnegie. For the most part, the younger scientists who joined him did not expand upon his most suggestive ideas. The Clementsian research group did not survive its leader, and later in his career Clements's thinking became increasingly idiosyncratic; the development of his older ideas came primarily from the work of scientists unassociated with the Carnegie Institution. Thus the ambiguity arises: Clementsian ecology was influential, but, although criticized for errors in his thought, he rarely receives credit for later innovations that sprang from his original ideas.

Clements and the Carnegie Institution

The Carnegie Institution of Washington was established in 1902 with an endowment of $10 million from steel magnate Andrew Carnegie.[5]

However, compared to the Rockefeller Institute, the other great supporter of American science during this period, the funding policies of the Carnegie Institution were rather nebulous. During the early decades of its existence the institution supported a wide variety of research projects, and there was considerable debate within the institution over the relative merits of various intellectual fields.[6] One field that received early and enduring support was botany, beginning with the establishment of a desert laboratory near Tucson, Arizona, in 1903.[7] During the next several decades botanical research supported by the Carnegie Institution expanded into a number of other areas including evolution, physiology and biochemistry, ecology, genetics, taxonomy, and paleontology.

During this formative period in its history Frederic Clements became associated with the Carnegie Institution. In 1903 he approached the institution for financial support with a request for eight hundred dollars to purchase instruments for his ecological studies on Pikes Peak.[8] Frederick Coville, a research associate at the Desert Laboratory who reviewed the proposal, was impressed with both Clements's proposed research and his qualifications as a scientist. However, citing the uncertainty in the institution's plans for botanical research he recommended deferring the award. "It is believed that the application should ultimately be granted," Coville wrote, "but that it would be advisable to defer the grant until a plan has been adopted by the Institution for the coordination of the researches bearing on this [ecological] subject."[9] Not until nearly a decade later did the Carnegie Institution begin to support Clements's ecological research. During the intervening years he established an international reputation with the publication of *Research Methods in Ecology* (1905) and was appointed chairman of the botany department at the University of Minnesota. In 1913 Clements took a six-month leave of absence to work at the desert laboratory. During the next three years the Carnegie Institution supported his research with small grants, and following the completion of his greatest work, *Plant Succession* (1916), Clements was appointed full-time research associate, a position that he held until his retirement in 1941.[10]

Clements joined the Carnegie Institution at a propitious time for program building. Not only did botanical research expand there throughout the 1920s, but the institution was also attempting to organize this research around evolution, development, and physiology—precisely the major themes of Clementsian ecology. Clements seemed to possess the characteristics necessary to play a central role in this reorganized botanical program. Under the presidency of Robert

S. Woodward (1904–1920) the funding policy of the Carnegie Institution of Washington had shifted away from small grants-in-aid and toward long-term funding of a few "exceptional men" whose research was central to well-established scientific fields.[11] Clements had established himself as a respected leader in the young discipline of ecology. *Plant Succession,* which was in the process of being published by the institution when Clements's research associateship was under discussion, promised to be a work of major significance. In his presidential report of 1916, Woodward extolled it as a "remarkable" and "profoundly instructive" book dealing with the central problems of ecology and evolutionary biology.[12] Woodward was not the only administrator impressed with Clements's intellectual abilities. Daniel Trembly MacDougal, director of the Department of Botanical Research and a man with a keen political sense, was also an enthusiastic backer. MacDougal had been instrumental in bringing Clements to the Carnegie, and his support seemed to bode well for Clements's future within the institution.

In his letter of appointment, Woodward wrote Clements that he wished to use the ecologist as an example of what it meant to be a full-time associate of the Carnegie Institution.[13] First, only scientists of the highest caliber would be considered for such positions. Second, in terms of salary and security, the institution was willing to offer its associates benefits equal to those enjoyed by full professors at leading universities. Perhaps most important for Clements's ambitious plans for the future, the letter suggested that his status within the institution would be equal to that of the director of a research department; thus, Clements would be able to hire a team of scientists to work under his direction. For a decade after Clements joined the institution his research group in ecology continued to enjoy this informal, quasi-departmental status.

Clements realized, however, that without formal departmental status there was a danger that he would lose control over ecological research, a fear borne out by later events. Therefore, almost immediately after joining the Carnegie Institution, Clements began lobbying for a separate department of ecology. In a memorandum to Woodward written in 1919, he outlined a comprehensive scheme for such a department. This outline faithfully reflected the conceptual framework presented in *Research Methods in Ecology* fifteen years before, with the notable addition of animal ecology as an important element in the program. Clements would direct this program in "pan-ecology" or "bio-ecology." Subdivided into four areas, each had its own permanent research associate: *experimental taxonomy* would investigate the evolutionary relationships among species; *experimental ecology* would study physiological adaptations in individual plants; *experimental vegetation* would extend Clements's earlier studies on the dynamics of plant communities; and *experimental biology* would investigate the ecological interactions between plants and animals. To carry out this ambitious research program Clements asked for four primary investigators, two full-time assistants, four or five part-time assistants, a secretary, and an annual budget of $34,000.[14] At a time when automobiles could be purchased for less than $500, this was no small sum. Clements eventually received most of his request, but Woodward was decidedly cool to his idea of a separate department for ecology. The economic uncertainty of the immediate post–World War I period was an important reason for Woodward's reticence, as perhaps was the fact that he was planning to retire as president. There were other important reasons for not supporting Clements's proposed department, reasons that Woodward did not articulate. One was the conservatism of the Carnegie Institution. In general, the institution supported research in well-established fields and avoided committing itself to novel ventures.[15] Clements's ecological research was acceptable within the broader intellectual context of botany, but as an independent field ecology was still struggling to establish an identity. An equally important reason for rejecting Clements's proposal was opposition from other biologists at the Carnegie Institution.

During the 1920s the Carnegie Institution was an intensely competitive arena, with many ambitious scientists jockeying for influence and control over institutional resources. Those less adept at internal politics could find their status within the institution greatly diminished. In her biography of Forrest Shreve, Janice Bowers has carefully documented the professional stumbling blocks that even talented scientists faced at the Carnegie. Shreve was a gifted botanist but not an astute "biopolitician."[16] Consequently his prestige within the institution suffered, and eventually his research program was eliminated. Clements's case was somewhat different: initially he had tremendous respect within the institution, and he was extremely ambitious. Although much more successful than Shreve, Clements's position within the institution also eroded over the years, a decline explained by the interplay between personalities and intellectual agendas.

Photographs of Clements invariably show an intensely serious individual (figure 2). Ecology was an obsession, and by midcareer the pressures of research had driven him to physical and emotional exhaustion.[17] Puritanical in personal habits, he abstained from tobacco and alcohol and was distressed by colleagues who did not. He was aloof, arrogant, and intellectually inflexible. Although flashes of humor occasionally appear in his correspondence, they seem strained and out of character. Even in Edith Clements's light-hearted memoir, *Adventures in Ecology* (1960), her late husband consistently comes off as a cold fish.

Character flaws do not preclude scientific creativity, but they can be professionally detrimental. Many scientists found Clements a difficult man to work with, and this affected his ability to direct a successful research group. Even before joining the Carnegie Institution, his prickly personality had almost derailed his career. Clements's first important professional break came in 1907 when he was appointed chairman of the Department of Botany at the University of Minnesota. Previously he had been teaching at his alma mater, the University of Nebraska. Clements was apparently not popular at Minnesota, and initial opposition to his appointment there was intense. Opponents presented the president and board of regents with a long list of complaints about the candidate's personality. These ranged from serious—that Clements was difficult to get along with and would lead to disharmony within the department—to scurrilous—insinuations that he and Edith Clements were not legally married. Critics suggested that he was being highly recommended only because the University of Nebraska wanted to get rid of him. Charles Bessey, Clements's mentor at the University of Nebraska, was sufficiently concerned about these charges that he composed a five-page letter rebutting each complaint against his former student.[18] When finally offered the position after a protracted search, Clements acknowledged that he was fortunate to get it, and he hoped that the department would "sooner or later feel the same way about their side of the bargain."[19] Clements chaired the department for a somewhat stormy decade, and the difficulties that his appointment faced at the University of Minnesota foreshadowed problems that he encountered later at the Carnegie Institution.

As botanical research at the Carnegie Institution expanded and diversified, Woodward and his successor as president, John C. Merriam (1920–1938), attempted to reorganize the program in a way that encouraged cooperation among researchers and simplified the administration of botanical work being carried out at several laboratories.[20] Eventually this led to the formation of a Division of Plant Biology with its headquarters on the campus of Stanford University.[21] However, the establishment of this new research unit came only after a decade of debate and negotiation within the Institution. Throughout the 1920s, Clements continued to campaign for a department of ecology, but other suggestions for reorganizing botanical research at the Carnegie Institution were also presented, some in direct opposition to Clements's ideas. Not only was this opposition partly personal and political, but it also reflected sharply divergent perspectives on the future course of botanical research.

Throughout the process of reorganization the administration placed great emphasis upon the necessity of coordination, cooperation, and efficiency in scientific research.[22] Implementing such lofty goals, however, among a diverse group of independent scientists was a major problem. Clements, with his domineering personality and single-minded commitment to ecology, was unlikely to encourage cooperation from other specialized botanists. From the beginning of his career, Clements had conceived of ecology as "the dominant theme in the study of plants, indeed, as the central and vital part of botany."[23] This grandiose view of ecology continued to inform Clements's plans for reorganization during the 1920s. In Clements's later schemes, all botanical research at the Carnegie Institution would cohere around an ecological theme, and other researchers appeared to be little more than subordinates in Clements's larger ecological research program.[24] Under any circumstances such a centralized research program could not have been very appealing to other prominent botanists at the Carnegie. The plan was even less appealing given the controversial nature of some aspects of his research, particularly as it applied to physiology and genetics. Indeed, even without his abrasive personality, Clements probably would have come into conflict with the Carnegie physiologists and geneticists for purely intellectual reasons.

Clementsian ecology, first outlined in *Research Methods in Ecology* (1905), was based upon the premise that both individuals and communities are organisms, and as such they share fundamental physiological properties. As a broad approach to ecological research, this physiological perspective was extremely suggestive, but many of Clements's specific claims about physiological cause and effect were naive. Traditional laboratory physiologists ridiculed him. In an extremely negative review, Charles Barnes, a physiologist at the University of Chicago, decried Clements's "vague explanations" and his "invalid reasoning."[25] Burton Livingston, a physiological ecologist at the Johns Hopkins University and a research associate at the Carnegie, dismissed *Research Methods* as "that *awful* book."[26] The book fared better with two British critics: the ecologist Arthur Tansley and the physiologist F. F. Blackman.[27] Although they wrote a long, generally laudatory review of *Research Methods,* correspondence suggests that the two reviewers were not equally enthusiastic about the book.

In a letter to Tansley, Blackman admitted that Clements's broad physiological approach to ecology was intriguing, but he was "repelled" by the details.[28] Criticism of his physiological approach to ecology followed Clements to the Carnegie. H. A. Spoehr, who eventually became the director of the new Department of Plant Biology in 1929, shared the disdain of other physiologists toward Clements's work. Spoehr was just as ambitious as Clements, and their relationship was generally hostile. Through private correspondence with President Merriam, Spoehr criticized the physiological work done in Clements's laboratory in an attempt to undermine the credibility of his research program.[29]

Clements's attitude toward genetics and his relationship with the geneticists at the Carnegie Institution was more complex. From the beginning, he believed in the inheritance of acquired traits and was convinced that speciation sometimes occurred by this process within a few generations.[30] Within the context of early twentieth-century botany this claim was controversial but legitimate, and inheritance of acquired traits fitted neatly into Clements's plans for an ecology based upon experimentation. By manipulating the environments of plants the botanist might be able to mimic the evolutionary process. This experimental taxonomy was a major part of his research program, and it increasingly dominated Clements's attention later in his career. During the early years of the twentieth century, such evolutionary views were within the mainstream of biological opinion, and the Carnegie Institution had a tradition of supporting research on the inheritance of acquired traits. Daniel Trembly MacDougal, who established the botany program at the Carnegie, was a prominent neo-Lamarckian. Even as late as 1925 other botanists associated with the institution, including the geneticists E. B. Babcock and A. F. Blakeslee, were willing to concede that Clements's experimental taxonomy was legitimate, perhaps even important, scientific research.[31] For example, in a long letter to President Merriam, Blakeslee wrote:

> The problem of the experimental induction of genetic change (mutations, heritable variations) in other words of experimental evolution is an alluring one to all biologists + has been the definite aim of the geneticist as indicated by the name of our Station at Cold Spring Harbor [Station for Experimental Evolution]. It must be admitted, however, that despite the enormous amount of thought + experiment on the problem, the most of which [*sic*] is unpublished, there is no single case during the last quarter of a century accepted by geneticists as an induction of a strictly genetic change.[32]

Blakeslee outlined the rigorous experimental conditions that needed to be met before geneticists would accept the validity of Clements's neo-Lamarckian claims. Despite his skepticism, however, Blakeslee expressed a willingness to cooperate with Clements on evolutionary studies.[33]

During the period of reorganization at the Carnegie Institution, Clements's commitment to neo-Lamarckism began to harden, and probably concern over his declining influence within the institution (discussed below) contributed to this change in his attitude. In August 1928 Clements wrote Merriam,

> It has now become possible to convert several Linnean species into each other, histologically as well as morphologically, and I need your judgment and advice as to kinds of evidence and the best methods of presentation. Both technique and results have reached the point where it should be possible to set the stage for a comprehensive demonstration within the next two or three years.[34]

Given the earlier discussions within the institution about experimental evolution, Merriam was greatly intrigued by Clements's claim, and he suggested that Clements present his preliminary findings before a conference of Carnegie biologists.[35] Whatever his motivation for writing to Merriam, Clements unfortunately did not have the data to back up his claim, and thus he had placed himself in an extremely awkward position. He was forced to delay presenting his results, a tactic that he continued until the end of his life. Perhaps anxiety over his position within the institution contributed to his increasingly anachronistic ideas about evolution. By the end of the 1920s he realized that his prestige within the Carnegie had seriously eroded, and he voiced concerns about the political maneuvering and intrigue surrounding the reorganization of botanical research at the institution.[36] Whatever the reason, Clements's experimental studies on the inheritance of acquired traits became an embarrassment for the Carnegie Institution, and after his death the Institution refused to publish the results of this research.[37]

The shifting balance of power within the Carnegie Institution during the 1920s is perhaps best illustrated by the changing relationship between Clements and his colleague, Harvey Monroe Hall. Hall was a professor of botany and curator of the herbarium at the University of California when he joined Clements's research group at the Carnegie Institution in 1919.[38] During the next decade the two men collaborated on transplant experiments, a venture that led to an influential but controversial book, *The Phylogenetic Method in Taxonomy.*[39] The personal relationship between Clements and Hall was cordial, but their working relationship was never particularly close. They did some field research together, but Hall continued to work primarily at Berkeley, while Clements did most of his experimental transplants on Pikes Peak. Intellectually, too, the experimental taxonomists diverged. Hall, believing that experimental taxonomy was a method for studying evolutionary patterns and mechanisms in general, never embraced Clements's neo-Lamarckian views. Furthermore, Hall was much more willing than Clements to expand experimental taxonomy to include genetics and cytology, and, unlike Clements, he developed a close relationship with the geneticists at the Carnegie Institution. One of Hall's closest friends was the geneticist E. B. Babcock. Beginning in the early 1920s, while Hall was still a member of Clements's research group, he and Babcock began exploring the possibility of combining ecological, taxonomic, and genetic methods to study the evolutionary relationships among plant species.[40]

Hall's synthetic approach to botanical research and his ability to work closely with a variety of other specialists were apparently exactly what the administration of the Carnegie Institution had in mind when they called for cooperative research. In contrast to Clements's idea of botanical research dominated by his own ecological research program, Hall's plan called for a much looser collaboration among botanists; independent researchers would continue their individual projects, but they would meet periodically in a "research conference" to explore areas of common interest.[41] This plan did not become the exact blueprint for the reorganization of botanical research, but Hall did become increasingly influential within the Carnegie Institution. Beginning in 1926, control of experimental taxonomy began shifting from Clements to Hall, and after 1928 Hall directed an independent research group that eventually refuted Clements's neo-Lamarckian claims.[42] Perhaps more important, President Merriam appointed Hall, Spoehr, and Blakeslee to an advisory committee on the reorganization of botanical research. Clements was excluded from the decision-making process, and the recommendations of the committee undercut his authority. Despite protests from Clements, the headquarters for the new Department of Plant Biology was built on the campus of Stanford University, not at Clements's winter laboratory at Santa Barbara. Spoehr, who held Clements in low esteem, was appointed chairman. Clements's research program, which had enjoyed quasi-departmental status for nearly a decade, was now simply one unit within the larger Division of Plant Biology. The budget for ecological research was now under Spoehr's control, an arrangement Clements bitterly resented.[43] An important part of the Clementsian research program, experimental taxonomy, was completely removed from his control, and during the 1930s both the scope and the staffing of his ecological research were further eroded.[44] By the time of his retirement in 1941, Clements's program in ecology amounted to little more than his studies on inheritance of acquired traits.[45]

Building a Research Group

Political skirmishes notwithstanding, the Carnegie Institution provided Clements with an enviable set of resources for building a dynamic research group. Financial support for his program was modest in comparison to some larger biological departments in the institution, but few ecologists during the 1920s were supported as lavishly as Clements. Within the restrictions set by his budget, Clements was completely free to build a research group as he desired.[46] The number of biologists working with Clements varied, but during the 1920s the research group was generally composed of four or five scientists working under his direction. Being a research associate at the Carnegie also allowed Clements to devote all his efforts to research. As his wife put it, after 1917 Clements was an "escaped professor."[47] The associateship freed Clements from the burdens of teaching, but it did not isolate him from academia. Students from the University of Nebraska could earn academic credit for fieldwork done at Clements's alpine laboratory in Colorado, and a steady stream of young scientists from other universities visited the facility.[48] From the beginning, Clements recognized that the prestige of a research program depended upon the "constant output" of articles and books.[49] He was a prolific writer, and his publications undoubtedly would have been influential under any circumstances. But the Carnegie Institution provided Clements with various outlets for publishing the results of his research. In particular, Clements effectively used the institution's monograph series to present extensive, detailed reports of his various research projects. *Plant Succession* (1916), his most influential work, was part of the series. And during the decade of the 1920s the Carnegie published seven other large monographs written by Clements and his coworkers.[50]

In a competitive scientific environment, Clements had an extraordinary set of potential resources for developing and disseminating his ecological ideas. Initially, these resources were used effectively, and much of Clements's influence on ecology can be attributed to the institutional niche that he exploited at the Carnegie during the 1920s. Owing to the interaction of a number of factors only partially under his control, Clements, however, was unable to create a dynamic research group that could survive after his retirement.

One key ingredient for the success of a research group is effective leadership. A leader must be capable of recruiting subordinates who are not only sufficiently committed to the leader's research to remain loyal to the program but also sufficiently independent to further develop the leader's ideas, produce conceptual or methodological innovations, and expand the research program into new areas. It may be difficult to strike the right balance between loyalty and intellectual independence; not surprisingly, even very productive research groups may be rather volatile.[51]

Clements was quite successful in attracting devoted younger biologists to his research group, some of whom stayed with him throughout his career. But given his domineering personality, Clements's coworkers found it difficult to develop independent lines of research within the broader context of Clementsian ecology. Group members seemed to be faced with only two options: subordinate their research completely, or break with the team. John Weaver and Harvey Monroe Hall, two examples of successful biologists who worked with Clements, illustrate this dilemma.[52]

John Weaver was not a great scientist, but he had a long and productive career as a botanist at the University of Nebraska. He served as president of the Ecological Society of America, and at the time of his death in 1966, he was recognized as the foremost authority on the ecology of North American grasslands.[53] Weaver was one of several Ph.D.s trained by Clements at the University of Minnesota. After completing his degree in 1917, Weaver joined the faculty at Nebraska, and during the summers he worked as a member of Clements's research group.

Weaver's major professional problem was his inability to successfully sever his student-professor relationship with Clements. Throughout their correspondence, including letters written two decades after Weaver completed his Ph.D., Clements's salutation was "Dear Weaver" while Weaver invariably opened his letters with a more respectful "Dear Dr. Clements."[54] These very different salutations indicate the subordinate position that Weaver continued to hold vis-à-vis his former teacher. From Clements's point of view, Weaver's research was simply part of the larger "organic whole" of the Clementsian research program.[55] Not only did he dictate the substance of

Weaver's work, which continued Clements's earliest interests in the ecology of the Nebraska prairies, but he also meddled in the younger scientist's professional life. On two occasions, Clements pressured Weaver to withdraw manuscripts that had been accepted for publication in the *Botanical Gazette*.[56] The reason was transparently obvious; Clements wanted Weaver's work to appear in Carnegie publications, thereby enhancing the status of his ecology program within the institution.

Clements originally planned to have Weaver become a full-time member of his research group, an offer that Weaver ultimately rejected.[57] Weaver's reticence was based upon several factors, including the financial security, access to graduate students, and equipment that his position at the University of Nebraska provided. But Weaver may also have been aware that to advance professionally he needed to free himself, however imperfectly, from Clements's domineering personality. Looking at Weaver's career, we see a scientist who was highly committed to Clementsian ecology; yet he did not have the independence to expand significantly the scope of this research tradition. Though competent, and even technically innovative, Weaver's research during the 1920s and 1930s remained thoroughly wedded to Clements's early views on plant ecology.

The experience of Harvey Monroe Hall provides an interesting contrast with that of Weaver. Unlike most other members of Clements's research group, Hall, having earned his Ph.D. at the University of California, was an outsider with no ties to either Nebraska or Minnesota. Hall and Clements were the same age, and by the time the Californian joined the team in 1919 he had already established his reputation as a plant taxonomist. To a much greater extent than any other member of the group, Hall was Clements's coequal. Significantly, Hall was the only member of the team to appear as senior author on a major publication written with Clements.

Hall left the botany department at the University of California to join Clements's research group, but only after protracted negotiations. Like Weaver, he was concerned about losing the security of a faculty position. In most cases, the Carnegie Institution made no formal commitments to Clements's coworkers; they were hired by Clements and served at his pleasure. Presumably, if Clements were to leave, the institution would not continue supporting the research group. But part of Hall's reticence was also intellectual; he and Clements approached scientific research very differently. "You are naturally radical and ready to take chances," Hall wrote Clements in 1918. "I am naturally conservative and want to carry the public with us. You are so anxious to get our results into immediate use that the early issuance of popular manuals is an important part of your program; I also want to serve in this manner but anticipate more keenly the intensive and detailed systematic studies in herbarium, garden, and field."[58] Hall was concerned as much with substance as with style, and throughout his life he remained noncommittal toward many of Clements's theoretical claims. In particular, he distanced himself from Clements's enthusiastic neo-Lamarckism.[59]

In contrast to Weaver and other members of Clements's research group, Hall maintained a high degree of scientific independence. He continued to live and work near Berkeley, and he forged strong ties with geneticists during the years when Clements was beginning to alienate them. Significantly, however, Hall did expand part of the Clementsian research program. He took one of Clements's earliest and most fruitful ideas, experimental taxonomy, and further developed it in a novel way. Clements had envisioned this line of research as a hybrid between ecology and taxonomy, using transplant experiments to determine phylogenetic relationships among species. Hall greatly expanded this interdisciplinary area of research to include not only ecology but also genetics and cytology. But Hall's innovation came at Clements's expense: During the administrative shake-up of 1928, research in experimental taxonomy was removed from Clements's control and placed under Hall's supervision. This was the only part of Clements's original research program that survived after his retirement in 1941. Ironically, the new team that Hall assembled beginning in the late 1920s—Jens Clausen, David Keck, and William Hiesey—effectively disproved Clements's neo-Lamarckian claims.[60]

The cases of John Weaver and Harvey Hall illustrate an important weakness in Clements's program of research. Clements built a productive research group, but this team did not successfully develop many key concepts that Clements had suggested in *Research Methods in Ecology* (1905). Most members of the group—Weaver, Herbert Hanson, Frances Long, and Glenn Goldsmith—worked on rather narrow empirical problems.[61] None of these biologists used Clements's conceptual framework as a foundation for developing bold, innovative ecological theories. Hall, who successfully exploited one of Clements's broader theoretical claims, did so in a manner detrimental to Clements's program, both institutionally and intellectually. Ironically, Clements's most suggestive claims—that the community is a kind of organism with its own distinctive physiology and that succession is a developmental process—remained relatively undeveloped. These ideas did not disappear when Clements died. Their further elaboration, however, came not from his students or the members of his research group, but from ecologists unaffiliated with the Carnegie Institution of Washington.

Clements's Influence on Modern Ecology

Critics have often claimed that Clements's influence was so pervasive that it amounted to a kind of intellectual stranglehold on ecology. Only after his death, it is argued, did ecologists free themselves from the misguided Clementsian paradigm.[62] This idea of the rise and fall of Clementsian orthodoxy is historical myth. During his lifetime, his ideas were the subject of vituperative attacks, particularly by some European ecologists. Even in the United States and Britain, supposedly the strongholds of Clementsian ecology, most of Clements's specific claims about communities and succession were challenged by prominent ecologists.[63] Among the first rank of Anglo-American ecologists it is difficult to find a single individual who might accurately be characterized as an orthodox Clementsian. Even within his own institution Clements was an embattled biologist. This is not to claim that he was uninfluential. To the contrary, Clements's ideas formed an important part of the foundation for ecosystem ecology. But his modern critics have distorted the historical context within which these formative ideas developed. The true significance of Clements's work was not that his ideas were accepted as a kind of orthodoxy, but that other ecologists considered them sufficiently important to criticize, modify, and use as the basis for further work.

Ironically, the significance of Clements's work is most apparent in his erroneous views. Before considering these views, it may be useful to consider a taxonomy of scientific error. Scientists are frequently wrong, and their errors, once discovered, often have little impact on the future course of science. But scientific errors may also play a more positive role. By vociferously arguing for an alternative explanation, even if it turns out to be wrong, a scientist may force opponents to clarify ambiguities in their thinking and make their assumptions more explicit. Richard Goldschmidt, who was wrong more often than he was right, may have played this role of gadfly in the history of genetics.[64] The case of V. C. Wynne-Edwards and his theory of group selection is another excellent example. Although quickly rejected, this theory stimulated considerable discussion of evolutionary mechanisms, and thus it played an important role in the emergence of evolutionary ecology during the 1960s.[65] A different situation arises when a scientist proposes a broad, superficially appealing generalization that can be shown to be false in the details. Despite serious problems the idea may stimulate further research, and a modified version may eventually become widely accepted. The case of Alfred Wegener's theory of continental drift is a notable example of a problematic concept that later gained acceptance.[66] Conversely, a broad generalization that initially appears plausible may later be discredited. Yet the controversy that leads to the rejection of the concept may, in itself, stimulate the development of an area of research. J. C. Willis's theory of "age and area" in plant geography may be an example of this type of fruitful scientific error.[67]

Clements's mistaken views fall into each of these four broad categories. The results of his research on inheritance of acquired traits were virtually ignored by other biologists by the time they were posthumously published in 1950. His unorthodox ideas had no discernable effect on the development of evolutionary biology. His claim that succession is always a linear process, progressing toward a uniform climax, was widely rejected during his lifetime. Because, however, this claim was presented in *Plant Succession,* the definitive work on the subject, it could not be ignored. His dogmatic assertions forced other ecologists to distinguish more carefully among various types of succession: primary and secondary, allogenic and autogenic. Clements's intransigence undoubtedly annoyed his contemporaries, but it seems likely that in the development of theories of succession Clements played an important role as gadfly. Similarly, his sweeping generalizations about climax vegetation later served as a foil for criticism.[68] Clements's most basic claim—that the community is a kind of organism with its own physiology—was erroneous in a more interesting way. The concept was never fully developed by either Clements or his coworkers; his discussions of the concept at the end of his career barely differed from the one presented in *Research Methods in Ecology* (1905). After World War II few, if any, ecologists accepted the strong version of Clements's organismal concept, but in a weaker form this concept was tremendously fruitful. As we shall see, the idea of the ecosystem was a direct descendent of the organismal concept. If later ecologists denied that ecosystems were organisms, then they continued to compare the physiology of organisms and the "physiology" of ecosystems. Thus Clements's broad physiological perspective and organismal concept played important roles in initiating a new line of research that greatly expanded after his death.

One might ask why Clements himself did not play a more active role in the later development of these ideas. Certainly part of the explanation lies in personality. Dogmatic and inflexible, he seemed psychologically incapable of acknowledging error or modifying his ideas. These personality traits were exacerbated by the political skirmishes that so dominated the later part of his career. Insecure in his embattled position, Clements was in no position to compromise with his critics. Another part of the answer was Clements's inability to assemble a dynamic research group. Successful research groups can be extremely effective in promoting the development of new lines of research. Using a suggestive analogy, David Hull claims that these groups are the intellectual equivalents of demes, the freely interbreeding populations of evolutionary biology.[69] The research group provides an informal social context within which ideas can be modified and transmitted more quickly than through the more formal channels of the extended scientific community. The group may also act as an efficient error eliminator. Nascent theories, even great ones, usually contain misconceptions, ambiguities, or unacknowledged assumptions. During the informal give-and-take between coworkers, ideas can be refined or modified before facing the scrutiny of outside scientists. Without the benefit of an effective research group a scientist may still be productive, but he is deprived of an important mechanism for perfecting, expanding, and perpetuating ideas. Such was the case with Clements. Many pieces of his theoretical system were useful, but he was unable to create a coherent intellectual tradition.

4

The Metabolic Imperative

From the viewpoint of a cat, mice are machines for converting plants into food.
—LAWRENCE B. SLOBODKIN, *Growth and Regulation of Animal Populations*

IN A REVIEW OF ONE of Frederic Clements's last major works, the Yale limnologist G. Evelyn Hutchinson remarked that if a community were an organism then it ought to have a form of metabolism.[1] This suggestive idea was never explored by Clements, and Hutchinson complained that Clementsian ecology had progressed little beyond the description and classification of communities. The criticism was not entirely fair. Whatever its shortcomings, Clementsian theory was oriented more toward process than classification.[2] Yet in his later years the great plant ecologist had done little to expand his organismal analogies, and to Hutchinson's aggressive intellect Clements's ideas appeared a bit shopworn.

By 1940, when Hutchinson's review was written, the contours of ecology were beginning to change. Like most of his contemporaries, Hutchinson refused to consider the community an organism. Yet the idea that the movement of energy and materials through the community was analogous to the metabolism of an organism intrigued him. Indeed, together with his close contemporary, Charles Elton, Hutchinson did much to place the study of this "community metabolism" at the heart of post–World War II ecology. In this chapter I consider the development of this physiological analogy and its theoretical ramifications at the hands of Elton and Hutchinson. Combined with the Clementsian notion of community development, community metabolism formed the intellectual core of what later became ecosystem ecology.

Charles Elton and Trophic Dynamics

Superficially, Charles Elton (figure 3) seems an unlikely historical candidate for stimulating the development of ecosystem ecology. In later life he showed little enthusiasm for either the new specialty or its emphasis on energy. Ecology, so he claimed, was "scientific natural history."[3] Elton delighted in detailed, descriptive field studies, and he found his inspiration in the Darwinian tradition of natural history rather than in the laboratory sciences. Nonetheless, through his insistence on the importance of feeding or trophic relationships, Elton laid the intellectual foundation for the study of energy flow in ecosystems.

Reminiscing on his childhood, Elton acknowledged an early interest in observing and collecting animals, an interest further stimulated by reading Darwin's *On the Origin of Species* when he was sixteen.[4] The transition from neophyte collector to mature scientific naturalist occurred when he entered Oxford shortly after the end of World War I. There he studied zoology under Julian Huxley, who was gaining an international reputation for his work on avian behavior. Perhaps more significantly, Elton accompanied Huxley on the Oxford University Spitsbergen Expedition, the first of three such journeys to the arctic that Elton made prior to writing his first book on ecology. These trips introduced Elton to field research, and they did much to shape his views on the nature of biological communities.[5]

Most earlier work on communities had been done by botanists, and Elton was willing to accept ecological succession as an important unifying concept. His views on the matter reflected the modified Clementsian scheme taught at Oxford by Arthur Tansley during the late 1920s. But community development was an inherently unattractive area of research for the zoologist. Unlike the botanist who could more easily study the slow changes in firmly rooted plant communities, the zoologist was forced to contend with constantly moving populations of often well-concealed animals. Zoologists, notably Victor Shelford, had completed some suggestive studies on succession in animal communities. In most cases, however, zoologists simply accepted the successional patterns previously determined for plant communities and then described the accompanying changes in the fauna. Elton had little enthusiasm for such derivative research.

In his audacious *Animal Ecology* (1927), no less remarkable because it was completed in three months when he was only twenty-six years old, Elton charted a new course for community ecology. "Animals are not always struggling for existence," Elton wrote, "but when they do begin, they spend the greater part of their lives eating. . . . Food is the burning question in animal society, and the whole structure and activities of the community are dependent upon questions of food-supply."[6] "Eating," a much more obvious phenomenon in animals than in plants, provided the zoologist with an important unifying concept. What unified animal communities—indeed, what made "community" a useful zoological concept—was the fact that within the community animals fed on other organisms. Whether one considered the microorganisms in a mouse's gut, the inhabitants of a small pond, or the fauna of an equatorial rain forest, the general pattern of feeding relationships was identical.[7]

The "general ground plan" of every animal community was an organized system of herbivores, carnivores, and scavengers.[8] But these traditional biological terms were too crude for the trophic analysis that Elton was creating. Through the course of evolution most animals had become specialists: an animal was not simply a carnivore; more likely it was a carnivore with rather restricted food requirements. Typically each species had an optimal food size. Potential prey that were too large could not be caught and destroyed; prey that were too small would not supply the nutritional requirements of the predator. This optimal range of food—not too big but not too small—determined the animal's niche in the community.[9] In describing the niche of an animal the ecologist was saying something about that species's status within the community, what the species was "doing" there. As Elton whimsically put it, "When an ecologist says 'there goes a badger' he should include in his thoughts some definite idea of the animal's place in the community to which it belongs, just as if he had said 'there goes the vicar.'"[10] Whatever Elton's opinion of the role of the clergy in English society, the role of an animal was always to eat or be eaten: "the 'niche' of an animal means its place in the biotic environment, *its relations to food and enemies.*"[11]

Recent critics have complained that the concept of niche is vague, little more than post hoc descriptions of the characteristics of animals. Thus, niche has little predictive value.[12] But in *Animal Ecology*, Elton suggested that this concept could play an important predictive role in ecological research. Communities shared a common ground plan, and even very different communities contained parallel niches. Therefore, given an unknown community, the ecologist ought to be able to predict the types of animals that existed there. Furthermore, it could be used as an important comparative tool for studying communities. Although taxonomically unrelated, species in two communities often played equivalent functional roles. For example, a common niche in the animal communities surrounding Oxford was that "filled by birds of prey which eat small mammals such as shrews and mice."[13] This niche was occupied by tawny owls in oak forests and by kestrels in grasslands.

Because animals tended to be specialists, feeding relationships in communities followed definite patterns. One could conceive of a *food chain* linking several niches in the community. This idea was not completely novel in 1927. It was a common practice among zoologists and fishery biologists to diagram feeding relationships with arrows leading from prey to predator. What made Elton's concept of food chain unique was the way he generalized it. The food chain represented the fundamental organizational structure or ground plan shared by all communities. And the passage of food through the chain of niches constituted the fundamental process that held communities together.

The "basic class" in the animal community was the herbivores, which converted plant material into animal tissue. A particular species of herbivore was consumed by certain carnivores, which in turn were eaten by other carnivores. The limits of such a food chain were set by the relative size and numbers of animals occupying each niche. Herbivores tended to be more abundant than the carnivores that ate them; the carnivores were living off the excess "margin" of the herbivore class. Carnivores that ate other carnivores tended to be both larger and fewer than their prey. In other words, as one moved up the food chain each level provided a smaller margin of food to the next. Thus at some point, the food chain reached its limit when the available food was too small to support an additional class of carnivores. This important relationship between predator and prey, which limited the length of food chains, Elton termed the *pyramid of numbers*.

In reality, this simple feeding scheme, represented by food chain diagrams and the pyramid of numbers, was complicated by several factors. Animals, in fact, had two sets of enemies: large carnivorous predators and small parasites. Field mice provided food for not only owls and kestrels but also fleas. Fleas were parasitized, in turn, by protozoans and bacteria. Thus, each level in a typical food chain was also the starting point for shorter, parasitic food chains. Other animals in the community lived by scavenging dead organic material, and a complete picture of feeding relations also required these scavenger or decomposer food chains. Finally, numerous food chains were linked together to form what Elton referred to as a *food cycle* and what a later generation of ecologists called a *food web*. Some animals—for example, copepods in a pond—were so numerous that they could support several different species of carnivores. Copepods were the "key industry" on which the entire economy of the community depended; many food chains radiated from this common supplier of food.

Though greatly simplified, Elton's idealized food cycle explained both the stability and the periodic disturbances found in nature. All populations had the potential for exponential increase. Left unchecked, this exponential increase would destroy the delicate balance that existed within communities. This occasionally happened, but usually populations remained relatively stable. Every animal population, Elton argued, had an optimum density. Because the environment was always fluctuating, the number of animals never remained constant, but populations tended to track a shifting optimum set by available food. Although food supply set the ultimate limitation on population growth, starvation was rarely the primary mechanism for controlling growth. In most cases, numbers began to decrease before the population faced starvation. The food cycle acted as a mechanism for this regulation. As the number of individuals increased, that population tended to become a more conspicuous target for predators. As numbers declined, predators shifted their eating habits and concentrated on another more plentiful prey species. Thus prey species were dependent upon predators for their own well-being; the predators acted as a kind of regulator maintaining the optimal numbers of their prey. Forty years earlier, Stephen Forbes had spoken of a "community of interest" between predator and prey; Elton shared this view. Speaking of the relationship between deer and their carnivorous "enemies," Elton noted that the predators "are in fact only hostile in a certain sense, in so far as they are enemies to individual deer; for the deer as a whole depend on them [predators] to preserve their optimum numbers and to prevent them [deer] from over-eating their food-supply."[14]

Elton presented the food cycle as a regulatory mechanism for maintaining equilibrium within the community. But important as it was, the food cycle was an imperfect regulator. Small herbivores sometimes reproduced so rapidly that predators could not control their numbers. For example, populations of lemmings, voles, and other microtine rodents periodically exploded. Migration served as one kind of safety valve for the community; as rodent populations reached high densities individuals dispersed to less populated areas. Another more important control mechanism was epidemic disease. The population cycles that so interested Elton were largely the result of periodic, unrestrained growth followed by catastrophic death, often the result of epidemic disease.[15] The food cycle, disease, migration, and other regulatory mechanisms maintained stability in the animal community, but this stability was a fragile equilibrium often subject to disruption.

Animal Communities and the Economy of Nature

Elton was a careful observer of nature who in later life spent twenty years completing an exhaustive study of animal communities near his home in Oxford.[16] He consistently emphasized the importance of such detailed field research as a foundation for ecology. But Elton was no naive empiricist. The growth of a coherent science depended heavily upon bold, general theories. Such theories, Elton claimed, were like carnivorous animals or powerful enzymes, capable of attacking and digesting apparently unrelated pieces of data.[17] The real genius of the young Elton, so obvious in *Animal Ecology*, was his ability to cut through the complexity of nature to uncover a more fundamental simplicity. Relying on a relatively small body of empirical evidence, Elton forged a set of basic ecological principles. These principles served as not only explanations but also guides for future empirical studies:

> The food-relations of animals are extremely complicated and form a very closely and intricately woven fabric—so elaborate that it is usually quite impossible to predict the precise effects of twitching one thread in the fabric. Simple treatment of the subject makes it possible to obtain a glimmering of the principles which underlie the superficial complication, although it must be clearly recognized that we know at present remarkably little about the whole matter.[18]

This sophisticated view of idealization provided a powerful rationale for studying ecology. Because, for all intents and purposes, every animal community shared a common economy or ground plan, the ecologist could investigate the simplest community and draw generalizations applicable to the most complex.[19]

Many of Elton's most fertile ideas were derived from his studies of very simple arctic communities. For example, his concept of the food cycle developed out of an early biological survey of a small arctic island.[20] In contrast to a complex woodland community in rural England, which might contain in excess of ten thousand species of animals, the impoverished fauna of Bear Island was represented by less than one hundred.[21] Elton's diagram of the food cycle of Bear Island (figure 4) included perhaps thirty-five animals, although terms such as "Diptera," "Protozoa," and "Marine Animals" obviously referred to more than one species. This elegant picture of the feeding relations on Bear Island did not emerge from years of detailed empirical study; he and his coworkers completed his survey of the animals inhabiting the island during a ten-day excursion when Elton was still an undergraduate in his early twenties. Truly remarkable was that the precocious biologist, using this brief investigation and a handful of other studies, arrived at fundamental principles of trophic dynamics—and that, at twenty-six years of age, he was able to convince other ecologists that they were valid.

A central theme running through *Animal Ecology* is the belief that communities are highly integrated, self-regulating entities. In discussing self-regulation Elton employed both organismal and mechanical metaphors, but the most striking analogy that Elton developed was one between animal communities and advanced industrial societies:

> Throughout this book I have used analogies between human and animal communities. These are simply intended as analogies and nothing more, but may also help to drive home the fact that animal interrelations, which after all form the more purely biological side of ecology, are very complicated, but at the same time subject to quite definite economic laws.[22]

Elton's economy of nature was based upon a kind of supply and demand system. Plants, which Elton barely considered, formed the raw materials for the herbivorous "key industries" of the community. The numbers of carnivores at the higher levels of each food chain were ultimately determined by the productivity of these key industries. Despite the potential for exponential increase, populations tended to oscillate about an optimum density set by the food supply. "This optimum number is not always the same and it is not always achieved," Elton admitted, "but in a broad way there is a tendency for all animals to strike some kind of mean between being too scarce and too abundant."[23] Borrowing an illustration from the sociologist Alexander Carr-Saunders, Elton compared the conditions faced by animal populations with the problem faced by the employer desiring an optimum number of workers: too few workers meant less than maximum productivity, while too many workers meant a decrease in profit.[24]

Nature's economy tended to be orderly and balanced, but like human industrial societies, animal communities were not completely free from violent and unpredictable events. An animal population occasionally increased beyond its optimal density, and this overpopulation could result in epidemic disease or starvation. The microtine population cycles that so interested him were obvious examples that Elton used to illustrate this phenomenon. But even more striking cases of catastrophic population fluctuations could be found in noncyclic species. Among other examples, Elton pointed to the now classic story of the mule deer inhabiting the Kaibab Plateau of Northern Arizona.[25] Prior to 1906 the population of deer had been maintained at about 4,000 individuals by wolves, mountain lions, and other predators. Beginning in 1906 the federal government began an aggressive program of exterminating predators. By 1931 most of the deer's natural enemies had been trapped or shot: 781 mountain lions, 30 wolves, 4,889 coyotes, and 554 bobcats. Unrestrained by predation, the population of deer expanded to approximately 100,000, a number far in excess of what the habitat could support. Starvation resulted in the death by the thousands, and eventually the population dwindled to some 10,000 deer. This scientific account has been challenged,[26] but for Elton, writing in the 1920s, it illustrated two important ecological generalizations. In a dramatic way it demonstrated that animal populations were capable of not only violent fluctuations but also self-regulation of the animal community as a whole. Temporary changes in a single population might have far-reaching effects on the food cycle of the community, but generally a natural balance was eventually restored. "In fact," Elton concluded, "if several important key-industry species become suddenly very abundant or very scarce, the whole food-cycle may undergo considerable changes, if only temporarily. The various automatic balanced systems which exist will tend to bring the numbers, and therefore the food-habits, back in the long run to their original state."[27]

The idea that nature formed a kind of well-regulated industrial economy was the dominant theme in Elton's early writings, but he was not entirely comfortable with this rather mechanistic picture of animal communities. In fact, there was a certain conflict in Elton's ecology between the industrial analogy and a conviction that nature was indeterminate, that stochastic processes were constantly interfering with the machinery of nature. So while he could write so eloquently of the "automatic balanced systems" that regulated communities, he could almost simultaneously reject the notion that there was a "balance of nature." In a series of lectures presented only two years after the publication of *Animal Ecology*, Elton stated that although the idea of balance was popular among biologists, it was untrue. "'The balance of nature' does not exist," Elton declared, "and perhaps never has existed. The numbers of wild animals are constantly varying to a greater or less extent, and the variations are usually irregular in period and always irregular in amplitude."[28] Elton also questioned whether communities were machinelike:

> The simile of the clockwork mechanism is only true if we imagine that a large proportion of the cog-wheels have their own mainsprings, which do not unwind at a constant speed. There is also the difficulty that each wheel retains the right to arise and migrate and settle down in another clock, only to set up further trouble in its new home. Sometimes, a large number of wheels would arise and roll off in company, with no apparent object except to escape as quickly as possible from the uncomfortable confusion in which they had been living.[29]

This rather striking statement did not necessarily reflect an intellectual conversion after the publication of Elton's first book; indeed, one can find similar, though muted, allusions to the unpredictability of nature in *Animal Ecology*. Rather, the thematic differences in his two early books reflect a deep-seated intellectual tension that ran through Elton's ecology. The probable sources of this tension were diverse.

To a certain extent, Elton's problem appears inherent to ecology. The same intellectual dichotomy can be found in Darwin's discussion of the entangled bank, and it continues to be source of considerable controversy in ecology today.[30] Elton was simultaneously tackling problems in population dynamics and community metabolism. Several years later, G. Evelyn Hutchinson, who was also interested in both these areas of ecology, suggested that studying them required two quite different points of view.[31] Population biologists tended to take a *merological* perspective, focusing upon independent individuals and assuming that populational phenomena determined higher level community properties. In contrast to this bottom-up approach, other ecologists, particularly those who later studied ecosystems, took a *holological* approach by studying the flow of materials and energy through food webs without considering the individuals that made up the web. Hutchinson, an eclectic biologist, seemed capable of making the transition from one perspective to the other effortlessly. Most other ecologists have not been so adept, for the change in perspective seems to entail more than simply differing points of view. Ecological terms and concepts also change meanings. For example, from the merological perspective, populations are not usually seen as parts of a community in the sense that gears are parts of a clock. Rather, they are "parts" only in the sense that together they form an aggregation. From such a point of view a process such as competition is often viewed as no more than an interaction among independent individuals. From the holological perspective, however, the parts of a biological system are quite often visualized as pieces of a complex mechanical-organic entity. In such a context, competition may be viewed as a kind of cybernetic or quasi-physiological *function* that stabilizes the entire community. The conundrum posed by these two perspectives persists today, and it is perhaps not surprising that in his path-breaking early work, Elton was simultaneously attracted and repelled by the clock metaphor.[32]

To explain Elton's contrary views on the balance of nature, one might also consider his use of analogy. He may have considered these merely illustrative figures of speech, but the particular analogies that he chose are revealing. To create a "scientific natural history" Elton needed to formulate fundamental laws and principles, and he looked to human sociology and economics for models. But writing during the uncertain 1920s, Elton could not be too sanguine about social and economic harmony, balance, or optimality. Such notions ran counter to the prevailing skepticism that characterized the work of many young intellectuals during the post–World War I years.[33] In 1926 one might be guardedly optimistic that industrial capitalism was basically sound, but it would have been difficult to be too confident about the future. Throughout the decade Britain's industrial machine had been "throttled down."[34] Chronic high unemployment, a trade imbalance, and labor unrest plagued the British economy. Elton's own professional situation, teaching part-time and working as a temporary consultant for the Hudson Bay Company, was particularly insecure. When the Great Depression began, Elton lost his job with the fur-trading company and scraped by on small research grants.[35] Perhaps not surprisingly, the industrial society that Elton saw in nature, though basically stable, was at times subjected to unpredictable and violent disturbances.

Elton's contrary views on the balance of nature also reflect quite different objectives in writing his two early books: *Animal Ecology* (1927) and *Animal Ecology and Evolution* (1930). In *Animal Ecology* he sought to lay out the fundamental principles of community ecology, and not surprisingly the major theme running through the discussion is the regularity of nature. He intended *Animal Ecology and Evolution*, a very different book, to critique widely accepted evolutionary ideas. Elton was highly critical of some Darwinians for believing that natural selection always produces adaptation. He accused them of accepting a naive evolutionary version of perfect adaptation, a "gossamer" of reassuring beliefs that every structure and behavior is of use in the struggle for existence.[36] Elton seemed painfully aware that the themes developed earlier in his *Animal Ecology*—balance, optimality, and self-regulation—tended to support this adaptationist position. In *Animal Ecology and Evolution* he turned his back on these themes; in fact, he parodied them.[37] He emphasized, instead, the capricious side of nature; animals could never be perfectly adapted in a constantly changing world.

Elton's own evolutionary ideas were complex, and they provided yet another source of tension in his ecological writings. He considered himself an heir to the Darwinian tradition in natural history, and he believed that natural selection was an important evolutionary mechanism. Like many other ecologists of the period, Elton considered natural selection to be a general process operating at all levels of organization from cells to communities.[38] It also served as a "potent stimulus" for ecological research and a "coordinating principle" for organizing ecological data.[39] Without the digestive juice of natural selection, population studies produced only a mass of unassimilated data. Finally, natural selection was an important part of the "automatic balanced systems" that maintained communities in equilibrium. This commitment to natural selection notwithstanding, Elton recoiled from what he considered the deterministic implications of Darwinian theory. "The Darwinian theory," he wrote, "which I take to be mainly true as far as it goes, regards animals as stationary units, or at any rate units with regular ecological habits, and acted upon by a number of selective eliminating agents. Darwinism, in fact, implies the selection of helpless bundles of flesh by an environment which lacks any traces of biological purposiveness."[40]

Given this attitude it is perhaps not surprising that Elton accepted genetic drift as an important mechanism for nonadaptive evolution.[41] But his deep-seated philosophical commitment to free will also led him to accept less orthodox evolutionary explanations. He was convinced that animals have some control over their evolutionary destinies. Animals, Elton believed, possessed an innate "awareness of environmental harmony."[42] This was more than a simple mechanical response to environmental stimulus; even insects were capable of making conscious choices. Through the gradual evolutionary emergence of mental activity, animals came to rely increasingly upon purposive behaviors. In higher animals, such purposive behaviors, or what Elton referred to as "traditions," might rival natural selection in evolutionary significance. "It is clear, then," Elton argued, "that among the higher animals we can perceive a method of evolution along a mental plane, unconnected with the spread of new mutations, a method which leads on a small scale to the production of customs, cultures, and gregarious habits, similar to those found in man."[43]

Elton apparently took his ideas on tradition—as well as some of his views on population dynamics—from Alexander Carr-Saunders. In his book *The Population Problem,* the British sociologist drew a distinction between animal evolution controlled by purely biological processes and human evolution guided, in part, by cultural traditions.[44] Elton blurred this distinction and argued that animals, too, have primitive traditions. Many behavioral traits, particularly in birds and mammals, were not inherited in a genetic sense; they resulted from parents teaching their offspring. For example, a mother bear swats her cub when it plays with something dangerous. Having no genetic basis, Elton argued, such maternal behavior could not arise through natural selection; it must have evolved through a rudimentary form of cultural transmission.

In the end, Elton presented his readers a rather eclectic picture of the animal world:

> The real life of animals is therefore a compound of many things: fixed and predetermined limits impressed by the environment; the relations of the sexes; the survival of things that are useful; a certain free will in the matter of choosing between good and evil surroundings, accompanied by a great deal of movement; a fairly large amount of pure chance; and sometimes a growing stock of new ideas born out of contact with new situations—Predetermination, Sex, Materialism, Free Will, Destiny, Originality, and Tradition.[45]

Predetermination and free will—Elton could hardly have it both ways. This apparent contradiction was symptomatic of the delicate and unstable equilibrium that existed in Elton's early writings. He was simultaneously attracted and repelled by a nature governed by mechanical law. Nature might be a highly mechanized, industrial society subject to deterministic natural laws, but at the same time it was a more chaotic world where even lowly creatures possessed considerable freedom to make choices.

Perhaps wisely, Elton never attempted to reconcile these views in a grand synthesis. After 1930, he directed his energies toward detailed surveys of local communities and studies of microtine population cycles. Late in his career he expressed a certain impatience with general theory.[46] But during that brief intellectual efflorescence of the late 1920s, Elton created an enormously flexible and fruitful theoretical scheme. The concept of community developed by Frederic Clements and other botanists had been plagued with an inherent rigidity;

community always referred to the plants inhabiting this or that particular geographical area. For Elton, the community became a much more abstract concept: it could refer equally to the intestinal fauna within a mouse or the inhabitants of the coniferous forest stretching across northern Canada; in either case, the same ecological principles applied. Animals formed a community precisely because they were organized into definite feeding patterns. The concepts of niche, food chain, food cycle (later referred to as food web), and pyramid of numbers became formative elements in the embryonic study of trophic dynamics, research that would become a major growth area in post–World War II ecology.

G. Evelyn Hutchinson: An Embryo Ecologist and an Embryonic Ecology

Elton's *Animal Ecology* was an extraordinarily successful book. After its initial appearance in 1927 it was reprinted eight times, and a paperback edition first released in 1966 went through another three printings within five years. Although it had originally been aimed at a general audience, Elton's first book had a profound impact upon professional ecology. According to the British ecologist Amyan MacFadyen, it "has probably inspired more ecological research than any other work."[47] One ecologist influenced by reading *Animal Ecology* when it first appeared was G. Evelyn Hutchinson (figure 5), then a young instructor at the University of Witwatersrand in South Africa. Later in his career, Hutchinson referred to the book as "deeply fundamental" and recalled that "it proved a stimulus by showing me that what I wanted to do in biology was indeed a significant part of the science."[48] For more than half a century Hutchinson pursued many problems posed by Elton's early book, but Hutchinson's approach to these problems was strikingly different from that of the author of *Animal Ecology*.

Elton's nonmathematical approach placed significant limitations on the development of ecological theory. He had made an important step by abstracting the concept of community from specific geographical instances; however, there was a tension between this process of abstraction and Elton's commitment to "scientific natural history." In contrast to the elegant picture of the animal community presented in his first book, what emerged from Elton's later research was detailed description only loosely draped upon the earlier theoretical framework.

Hutchinson was not a pure theoretician. Like Elton, his research was a mixture of theoretical and empirical studies, including much work on the natural history of aquatic invertebrates. But Hutchinson was much more adept at the process of idealization. In contrast to Elton, whose nonmathematical ecology never strayed far from his descriptive natural history, Hutchinson was comfortable with the newer mathematical theory in ecology that began to develop during the late 1920s. Like the cyberneticians whom he admired, Hutchinson was quite willing, on occasion, to abandon Elton's natural history; for Hutchinson, nature was often cast as a "black box" whose inputs and outputs were the primary focus of study. This highly abstract approach to theory had a significant effect on the development of ecosystem ecology. Some factors that influenced Hutchinson's rather unique style of research can be gleaned from his biography.

Like Elton, who was two years his senior, Hutchinson grew up in an academic environment. Hutchinson's father, a chemist at Cambridge University, encouraged his son's boyhood interests in natural history and geology.[49] These informal scientific interests were reinforced when young Evelyn attended Gresham's School, one of the few English public schools to emphasize the sciences rather than the more traditional classical curriculum. Hutchinson always maintained an interest in the natural history of organisms. But unlike Elton, whose aversion to the physical sciences was undisguised, Hutchinson reserved central roles in his ecology for both chemistry and geology.

Charles Elton's ecological interests were set early in his undergraduate career at Oxford. In contrast, Hutchinson pursued a rather eclectic education at Cambridge. From boyhood he had become familiar with many scientists at the university. But apparently no one served the role of mentor in the way that Julian Huxley did for Elton. In his memoir, subtitled *Recollections of an Embryo Ecologist,* Hutchinson characterized his college experience as that of "a hunter and gatherer rather than that of someone settled in the industrious pursuit of intellectual agriculture."[50] This eclecticism became one hallmark of his career as a scientist.

After a year at the Stazione Zoologica in Naples and two years of teaching in South Africa, Hutchinson was appointed an instructor in the zoology department at Yale University in 1928. There a number of circumstances conspired to bring together the disparate strands of his early training, creating a unique approach to ecological research. These circumstances deserve extended consideration.

Biogeochemistry of a Connecticut Lake

Charles Elton had laid the theoretical groundwork for the development of trophic dynamics with his discussion of food chains, niches, ecological pyramids, and the food cycle. These concepts were fundamental to the study of a community metabolism, but in retrospect Elton had missed a crucial distinction between the flow of energy and the cycling of materials in ecological systems. In one of his earliest papers Elton discussed a "nitrogen cycle," a term he used interchangeably with food cycle. However, this nitrogen cycle is not the one familiar to the ecology student of today.[51] Implicit in Elton's diagram is the idea that both matter and energy move through the cycle (figure 3). But a distinction between the ways that matter and energy move is critical to modern ecology. In theory, matter can be transferred endlessly without any loss; it truly cycles. But because some energy is dissipated during each trophic transfer, energy flows *through* the system. In other words, ecological systems require a constant input of energy, not necessarily materials. One ought not be too critical of an intellectual pioneer for failing to see all the important implications of his work. But certainly one reason why Elton failed to distinguish explicitly between matter and energy was his antipathy toward the physical sciences. The modern understanding of material cycling and energy flow rests on not only biology but also geology and chemistry. In other words, it is the study of biogeochemistry. Hutchinson was much better prepared than Elton for this kind of interdisciplinary approach to ecology.

From boyhood, Hutchinson was attracted to geology and chemistry, but his mature interest in biogeochemistry was catalyzed by reading the works of Viktor M. Goldschmidt and Vladimir I. Vernadsky.[52] Goldschmidt was an important geochemist and a friend of Hutchinson's father. Hutchinson was introduced to Vernadsky's ideas by the Soviet theorist's son, a historian at Yale University.[53] Together, the younger Vernadsky and Hutchinson translated some of the father's writings and arranged to have them published in American journals. Vladimir Vernadsky's ideas were particularly important for the young limnologist. According to Vernadsky, life existed only in the *biosphere,* a thin layer composed of the upper regions of the oceans, the surface of the earth, and the lower regions of the atmosphere. Neither the term nor the idea was original, but Vernadsky's emphasis upon the cycling of chemical elements, a process that involved a close interrelationship between the living and nonliving components of the biosphere, made his work unique. Indeed, according to Vernadsky the major gases in the earth's atmosphere—oxygen, nitrogen, and carbon dioxide—were all of organic origin. This emphasis on living organisms—the belief that *geochemistry,* in fact, ought to be *biogeochemistry*—was a major innovation with important implications for ecology. Earlier ecologists, notably Frederic Clements, had considered the relationship between biological communities and the nonliving environment. But for Clements the relationship had been one of simple physical causes and biological effects. This simple mechanistic notion of causation was being replaced by a more complex interactive concept of ecological systems, and Hutchinson's ecological studies in biogeochemistry were pioneering contributions to this movement.

Hutchinson participated in theoretical debates concerning the movement of chemical elements through the biosphere as a whole, but these forays into global biogeochemistry rested on detailed investigations of chemical cycling on a smaller scale. For about a decade and a half beginning in the mid-1930s, Hutchinson and several of his students intensively studied the history, productivity, and biogeochemistry of a small, relatively self-contained system: Linsley Pond. The relationship between Hutchinson's pond studies and his theoretical discussions of global biogeochemistry was not unlike that between Charles Elton's early studies of arctic communities and his general theoretical discussions of trophic dynamics. Both ecologists were committed to the belief that generalizations derived from the simplest ecological systems could be extrapolated to very large, complex systems. According to Hutchinson, the self-regulatory mechanisms governing the biogeochemical processes of Linsley Pond were comparable to those operating in the biosphere as a whole.[54]

Linsley Pond, located on the outskirts of New Haven, is a small, nutrient-rich (eutrophic) lake with a surface area of about twenty-five acres and a maximum depth of about fifty feet. Beginning in 1935, Edward S. Deevey, one of Hutchinson's earliest graduate students, began studying the deep sediments that form the bottom of the lake.[55] Using a long boring device, Deevey removed earthen cores containing up to forty feet of sediment. The longest cores recorded nearly the entire history of the lake, beginning at the end of the last glaciation. Viewed under the microscope, samples from the core revealed the fossilized remains of tiny invertebrates, which provided a record of succession in the aquatic animal community. Identifying and counting pollen grains in the sediment revealed the successional changes in plant communities surrounding the lake and provided indirect evidence for climatic changes in the region. Hutchinson used samples from one core to analyze historical changes in the chemical composition of lake bottom.[56]

The record of organic material was particularly significant, for it provided a measurement of the productivity of the lake as a whole. In the very deepest samples the sediment was practically devoid of organic material, reflecting the relatively sterile condition of a newly formed glacial lake. Through time, as phosphorus- and nitrogen-containing compounds entered the lake in runoff, productivity increased. However, productivity did not increase in a simple, linear fashion. Once inorganic nutrients, particularly phosphorus, no longer acted as limiting factors for growth, productivity increased exponentially. This exponential rate of increase continued for only a relatively short period in geologic time. After the lake became eutrophic, its productivity remained essentially constant. This *trophic equilibrium* resulted primarily from the complex biogeochemical metabolism of the lake: the influx and efflux of nutrients, their cycling between water and sediment, and their movement through the food chains of the aquatic community. Trophic equilibrium was maintained until the lake, becoming a bog or marsh, ceased to be a lake.

Considering the sigmoid growth curve for productivity reflected in the historical record of sediments from Linsley Pond, Hutchinson remarked: "It is impossible to avoid qualitative comparison of this mode of development of the rate of organic production of sediment . . . to the growth curves of individual organisms and homogeneous populations."[57] Of course, there was an important difference between the development of an aquatic community and the growth of an individual organism or a population. In the latter two cases, there was a genetic continuity between either cells of the individual or generations of the population. This genetic continuity was responsible for the tendency for such systems to grow exponentially. Such a relationship did not exist within a community made up of many different, and often quite unrelated, species. Hutchinson suggested, however, that another mechanism might account for the exponential growth in productivity. During the course of succession there was a general tendency for organisms to modify the environment in such a way that it would support a greater quantity of living matter in the future. For example, nitrogen-fixing bacteria removed this important element from the atmosphere and made it available to other organisms. "Such a process," Hutchinson concluded, "would produce an effect essentially similar to that of the tendency to geometrical increase in genetically continuous systems."[58]

The sigmoid curve of productivity could be interpreted as a reflection of the overall growth of the aquatic community or "biocoenosis." A similar, though not identical, pattern of development seemed characteristic of certain parts of this organic whole. For example, by counting the microfossils of the crustacean *Bosmina* in various samples of the core, Edward Deevey discovered that the growth of populations of this invertebrate genus corresponded closely to the exponential phase of the productivity curve, though the rate of growth of *Bosmina* was greater. "In the language of the analogy between the organism and the biocoenosis," Deevey wrote, "this evidently means that a sort of allometric growth occurs, the rate of growth of the part differing from that of the whole, as the claw of the fiddler crab grows faster than the animal as a whole."[59]

These rather striking suggestions that aquatic communities are like organisms should not be taken to mean that Hutchinson and Deevey were working within some sort of Clementsian paradigm. The immediate source of the ideas was probably not Clements or any other ecologist. Rather, these ideas apparently were stimulated by discussions in a seminar led by the embryologist Ross Harrison.[60] During his early years at Yale (1930–1937) Hutchinson was responsible for teaching the undergraduate course in embryology. Therefore, he and his students may have been particularly receptive to such developmental analogies.

Analogies between embryonic development and ecological succession were useful for suggesting or clarifying ecological questions, but, although they took the idea of the community as a kind of "superorganism" just as seriously as Clements did, Hutchinson and Deevey were using the organismal analogies in a much more abstract, idealized fashion.[61] The truly important similarities between individual organisms and communities were at the level of mathematical equations representing energy transfer, material cycling, or growth of some biological parameter. Hutchinson, as noted at the beginning of this chapter, considered Clements's organismal analogy to be little more than a vague descriptive device. The study of succession had suffered because too much emphasis had been placed upon changes in the taxonomic composition of the community over time. These were superficial morphological changes. More fundamental, from Hutchinson's point of view, were the changes in community metabolism underlying succession. Clements had presented a physiological explanation of succession, but it was comparatively crude. By combining the biogeochemistry of Vernadsky, the trophic dynamics of Elton, and the use of mathematical models, Hutchinson and his students were creating a much more sophisticated explanation of the development and metabolism of aquatic communities.

Hutchinson's early work on the metabolism of lakes culminated in the work of his protégé Raymond Lindeman. Lindeman's contribution to the emergence of ecosystem ecology was so significant that I devote an entire chapter to his work. At this point suffice it to say that by 1940, Hutchinson and his students had developed a distinctive approach to limnology. They were working within a general organismal framework. Although compared to Clements, they used organismal analogies in a more flexible and sophisticated manner. These analogies provided a suggestive language for discussing ongoing processes, historical changes, equilibrium, and self-regulation within the aquatic community. More important, they served as heuristic devices for suggesting mathematical explanations for ecological processes. Succession was not ontogeny, as Clements had claimed, but an equation for allometric growth might provide a common explanation for both forms of development. This abstract, mathematical theorizing became increasingly important in Hutchinson's ecology after World War II.

Circular Causal Systems in Ecology

Most of Hutchinson's early research had centered around the study of lakes, but he had also become interested in more general theoretical problems in ecology. According to Hutchinson, ecology encompassed a variety of phenomena, ranging from purely biological processes to complex chemical and physical ones. Ecological problems between these two poles could not be broken up into discrete categories, but the Yale limnologist suggested that ecology could be conveniently characterized by two broad methodological approaches.[62] One, the *biogeochemical* mode, involved the interdisciplinary study of the movements of materials and energy through the biosphere. The second mode of ecological research was *biodemographic.* This approach was purely biological, studying numerical variations in the sizes of populations. Unifying the two approaches, according to Hutchinson, was the concept of negative feedback. Whether studying the phosphorus cycle of a lake in Connecticut or changes in the size of a population of aquatic invertebrates, the ecologist was studying self-correcting systems governed by negative feedback. This common property of ecological systems was not merely a vague, superficial similarity; there were *formal analogies* between the processes of population growth and biogeochemical cycling. In other words, regulatory mechanisms in both cases could be described by a common mathematical expression.

Hutchinson's most detailed discussion of negative feedback and formal analogy was delivered at a 1946 conference on "Teleological Mechanisms," one of a series of interdisciplinary meetings sponsored by the Josiah Macy Foundation. These conferences were designed to present the new field of cybernetics, the study of self-regulating systems, to the broader scientific community. Participants at the meetings included a diverse group of luminaries from the fields of mathematics, anthropology, social psychology, engineering, physiology, ecology, and philosophy.[63] Given the diversity of the participants, the discussions at the conferences were sometimes rather amorphous. Although some participants found the meetings fruitless, many considered them worthwhile.[64] For Hutchinson, the conferences provided a stimulating intellectual environment and an unusual opportunity to synthesize his ecological ideas.

The central ideas of cybernetics—system, self-regulation, feedback, oscillation, and time lag—were used by a diverse group of intellectuals, but exactly how the ideas were employed varied greatly.[65] The core group in the new field was composed of mathematicians and other mathematically sophisticated researchers drawn from fields such as computer science and physiology: Norbert Wiener, Arturo Rosenblueth, Julian Bigelow, John Von Neumann, Warren McCulloch, and Walter Pitts. These cyberneticians were attempting to construct an overarching mathematical theory to explain the behavior of organisms, machines, and other complex systems. In fact, from the perspective of cybernetic analysis, distinctions between machines, organisms, and even societies seemed to evaporate; all three were treated mathematically as "systems." Not everyone who was attracted to cybernetics possessed the mathematical skills needed to understand fully the theory being developed by the core group. For example, the anthropologist Gregory Bateson, a regular participant at the Macy Conferences, understood little mathematics, but cybernetics offered him a set of very general concepts that could be employed heuristically or metaphorically to explore social interactions.[66]

Hutchinson's use of cybernetics fell between these two extremes. Mathematically, Hutchinson was not in the same league with Wiener or Neumann, but, compared to Bateson, he was able to explore more fully the mathematical ideas developed by the cyberneticians. The Macy Conferences provided a stimulating intellectual environment within which he could apply these ideas to ecological problems. The concept of negative feedback appeared prominently in his conference paper "Circular Causal Systems in Ecology" and formal analogy, another idea central to cybernetics, formed the unifying theme of the paper. Hutchinson had employed formal analogy in his work for more than a decade. But the Macy Conference provided him with a venue for exploring, in much greater depth, its theoretical implications.

The logic of Hutchinson's "Circular Causal Systems in Ecology" combined these cybernetic concepts with the mathematical ecology created during the 1920s by Raymond Pearl, Alfred J. Lotka, and Vito Volterra.[67] The basic equation that Hutchinson employed was one originally formulated by the nineteenth-century mathematician François Pierre Verhulst and independently some seventy years later by Pearl. The Verhulst-Pearl equation describes the growth of a population within an environment containing limited resources. Using Hutchinson's notation

$$dN/dt = Nb(K - N)/K$$

where N was the population size, b represented the reproductive rate (*intrinsic rate of natural increase* in today's terminology), and K was what Hutchinson termed the "saturation level." This saturation level, today more commonly referred to as carrying capacity, is the maximum population sustainable by the environment. Without the term, $(K - N)/K$, the population grows exponentially. With the term, the population exhibits a sigmoid growth curve, approaching an asymptote at $N = K$. Biologically, the term $(K - N)/K$ reflects the fact that as a population increases it depletes resources in the environment. At least in theory, this depletion of resources reduces the rate of growth of the population. Thus, when the population is very small, $(K - N)/K$ is approximately 1, and population growth approaches the exponential growth described by $dN/dt = bN$. As the population becomes larger, $(K - N)/K$ becomes smaller, and the rate of growth continuously decreases. As $(K - N)/K$ approaches 0 growth almost ceases, and the population approaches, but never quite reaches, its saturation level (K).

Hutchinson claimed that the term $(K - N)/K$ constituted the mathematical expression for self-regulating mechanisms. Mathematically the term could be used to describe negative feedback, regardless of what biological form this feedback took. In other words, Hutchinson constructed a *formal analogy* among seemingly unrelated biological processes: individual growth, population growth, and historical changes in the productivity of an aquatic community. Clearly, the biological mechanisms causing these diverse phenomena had little in common. However, the formal analogy suggested a common method for explaining quite different ecological problems. Thus, it provided a theoretical bridge between the biogeochemical and biodemographic approaches to ecology.[68]

The simple logistic equation described the behavior of biological systems approaching a stable equilibrium. However, cybernetics also provided Hutchinson with engineering and physiological analogies for unstable systems. Cybernetics had developed largely out of the wartime problem of designing an automatic control system for anti-aircraft guns.[69] The purpose of such a device was not to aim the gun at the aircraft itself, but to direct the trajectory of the projectile toward some future position of the target. Information, in this case the difference between the present position of the gun and the future position of the target, was used to modulate the movement of the gun. Properly constructed this feedback mechanism would produce a series of dampened oscillations, swinging the gun in a smooth arc until it arrived at the proper firing position. However, if there were an appreciable delay in the feedback loop, the gun would oscillate wildly back and forth, never finding its target.

This mechanical failure in the anti-aircraft gun had a pathological analogy in neurophysiology. A patient suffering from purpose tremor is unable to successfully grasp objects because of uncontrollable oscillations of the hand.[70] Hutchinson believed that time lags in the self-regulatory mechanisms of populations might have similar effects. If there was very little lag, then oscillations would decrease as the population growth curve approached *K*. However, larger oscillations would appear if there was an appreciable time lag in the system:

$$dN_{(t)}/dt = bN_{(t)}(K - N_{(t-T)})/K$$

In this situation, rather than a smooth approach to *K*, the population would oscillate around the saturation level, alternately overshooting and falling below *K*. Such a situation might occur where the production of eggs required a significant period of time. The number of eggs released would be determined not by the population density at the time of laying (t), but rather the density at an earlier time ($t - T$) when the eggs began developing in the female's body. Like the anti-aircraft gun and the patient's hand, this population would hunt for, but never find, a stable equilibrium.

Oscillations, in any system, are potentially destabilizing. Dense populations are subject to catastrophic destruction by epidemics, and sparse populations risk extinction through random events. Therefore, Hutchinson reasoned, fluctuating populations tend to be replaced by their more stable neighbors. Natural selection, acting on various populations, improves self-regulating mechanisms by reducing time lags. As a result, most natural populations are in equilibrium most of the time.

Negative feedback mechanisms were important for not only regulating population size but also stabilizing communities. Hutchinson's discussion of predator-prey interactions is particularly intriguing for two reasons. He attempted to put Charles Elton's idea of "automatic balanced systems" on a firm mathematical footing, and in this section of the paper he most explicitly used cybernetic concepts to bridge the biodemographic and biogeochemical perspectives in ecology.

Elton believed that predation was one of the most important mechanisms for regulating animal communities. The size of prey populations was regulated by predators, and to some extent the growth of predator populations was checked by available food. During the 1920s, Alfred J. Lotka and Vito Volterra had described such a situation in mathematical terms.[71] The relationship between a predator and its prey could be described by a pair of differential equations. The rate of population growth of the prey was determined by its own natural rate of increase (b_1N_1) minus the rate at which prey were destroyed by predators. Assuming that the rate of destruction was proportional to the rate at which predators encountered prey ($p_1N_1N_2$), where p_1 was a constant that represented the efficiency of the predator, then:

$$dN_1/dt = N_1b_1 - p_1N_1N_2.$$

Assuming that the growth of the predator population was proportional to the rate at which prey were ingested ($p_2N_1N_2$) minus the natural death rate of predators (d_2N_2), then

$$dN_2/dt = p_2N_1N_2 - d_2N_2.$$

In such self-regulating systems, each population acted as a kind of mechanical governor, limiting the growth of the other population. Predators acted by influencing the death rate, but not the birth rate, of the prey. The prey acted by influencing the birth rate, but not the death rate, of the predators. Theoretically, the two populations oscillate slightly out of phase with one another.

Despite the theoretical predictions of the Lotka-Volterra model, few cases of predator-prey oscillations could be found in nature. Such oscillations could be induced in laboratory populations, but only under highly artificial conditions. Hutchinson, again, cited this negative evidence as an indication that natural selection had acted to perfect the self-regulatory machinery of the community. The oscillations inherent in the behavior of predator-prey systems had been minimized as the two populations coevolved.

The Lotka-Volterra equations could also be used to study biogeochemical cycles. A decade earlier, a little known Soviet geochemist, Vladimir Kostitzin, had used the equations to model a simple carbon cycle involving only plants and animals.[72] In this model, the population growth of animals was regulated by the photosynthetic production of plants, and, in turn, the population growth of the plants was regulated by the carbon dioxide liberated by the animals. In such a system, not only would the populations of plants and animals oscillate, but so too would the amount of carbon dioxide in the atmosphere. While very small oscillations in carbon dioxide levels might go undetected, Hutchinson argued that there was no evidence for significant atmospheric fluctuations. This suggested that the biosphere contained complex self-regulating mechanisms acting to maintain carbon dioxide at a constant level.[73]

The major component of this self-regulatory apparatus was the ocean, which acted as an "enormous shock-absorber" for carbon dioxide. The ability of the oceans to absorb and release carbon dioxide to the atmosphere was mediated by a system of chemical equilibria among calcium carbonate, calcium bicarbonate, and carbon dioxide:

$$CaCO_3 \longleftrightarrow Ca(HCO_3)_2 \longleftrightarrow CO_2.$$

Carbon entered the oceans primarily from rivers, which acted as a conveyor belt carrying large amounts of bicarbonate and smaller amounts of carbon dioxide. Carbon left the oceanic reservoirs primarily through precipitation of calcium carbonate on the ocean floor and diffusion of carbon dioxide between the ocean surface and the atmosphere. A number of slower geochemical processes were also involved in the carbon cycle, but the major features of this system were exchanges of carbon among the atmosphere, oceans, and biological communities. Because the oceans contained such vast reserves of carbon, they buffered the entire system against the oscillations in carbon dioxide predicted by Kostitzin's simple mathematical model.[74]

Hutchinson went to considerable lengths to justify the belief that biogeochemical cycles were self-correcting. In a section of his paper, "The Efficacy of Self-Regulating Mechanisms," he challenged the earlier claim by British geochemist G. S. Callendar that industrial combustion of fossil fuels was increasing the amount of carbon dioxide in the atmosphere. Carbon dioxide levels had increased in certain industrial regions, but, according to Hutchinson, this local phenomenon did not indicate a change in the gaseous content of the atmosphere as a whole. Furthermore, Hutchinson argued that local increases in carbon dioxide were more likely owing to deforestation in urban areas, rather than combustion of fossil fuels. He was confident that the self-regulating mechanisms of the biosphere were capable of correcting imbalances in carbon, whether natural or artificially produced.

Neither volcanos nor human industry were likely to alter the global levels of atmospheric carbon dioxide.[75]

One ought not be too critical of Hutchinson for failing to anticipate a later generation's concerns about the greenhouse effect, but his 1948 discussion of the carbon cycle is striking for other reasons. It reflects his deep commitment to the concept of equilibrium. Unlike Charles Elton, who was sometimes uncomfortable with the idea, Hutchinson was convinced that nature's self-regulatory mechanisms are capable of maintaining a balance even in the face of extreme perturbations. And unlike Elton, Hutchinson emphasized the role that natural selection played in perfecting this self-regulatory machinery and maintaining it in an optimal state. From Hutchinson's perspective the population cycles that so interested Elton were "derangements" from self-regulation, deviant phenomena that could only exist under unusual environmental circumstances.[76] Hutchinson's commitment to equilibrium was certainly not unique, although it is difficult to find another ecologist who believed so deeply in the balance of nature. More unusual about Hutchinson, however, was the way he attempted to place the concept of equilibrium on a firm mathematical foundation and tried to unify theoretical ecology by drawing formal analogies between very different kinds of self-regulating systems.

Building a Research School

Students and colleagues are often almost reverential in their admiration for the "Hutchinson mystique."[77] By the time he retired in 1971, some forty doctoral students had completed their dissertations under Hutchinson's direction, and his bibliography included nearly two hundred entries covering a broad array of biological topics. Many of his students played pivotal roles in shaping the ecology of the 1950s and 1960s. Hutchinson's success as an ecologist depended upon a number of factors: location, colleagues, and scientific style. Where one is located can have an important bearing on the success of a research program. As a young biologist in his late twenties, Hutchinson joined the faculty of a prestigious university, but a university with virtually no reputation in ecology. The only other ecologist at Yale during the 1930s was G. E. Nichols, whose interest was limited to botanical topics. The zoology department, devoid of ecologists, offered Hutchinson a relatively open niche for exploring unconventional problems and novel approaches to research without worrying too much about the reactions of established colleagues. Limnology,

Hutchinson wrote to his parents soon after arriving at Yale, was "the dumping ground for inferior off-scourings of the profession of zoologists."[78] In a zoology department with an established reputation in ecology, such a marginal area of research might well have been unattractive to an ambitious, young biologist. But with the encouragement of his sympathetic chairman, Ross Harrison, Hutchinson established a somewhat unorthodox research program that eventually became a major center of ecology in the United States.[79]

A great measure of Hutchinson's success was also owing to his ability to gather around him talented students and coworkers, associates who not only absorbed his ideas but modified and extended them as well. Even as a young unknown, Hutchinson was apparently able to attract and stimulate students. Gordon Riley, Hutchinson's first doctoral student, recalled: "Many students have come to Yale expressly to work with Evelyn Hutchinson. I did not. I had not heard of Evelyn before I came there. He was a young instructor—a nobody—but he was a great nobody, and the first month of his limnology course convinced me that my path lay in that direction."[80] Hutchinson's none too subtle displays of erudition irritated some students, but his enthusiasm for ecology was apparently infectious. His protégés—Raymond Lindeman, Howard Odum, and Frederick Smith—became key figures in the emergence of ecosystem ecology.

What apparently attracted many students to Hutchinson was his imaginative and innovative approach to theoretical ecology. Like most scientists, Hutchinson believed that theories should be potentially testable. However, unlike many scientists, Hutchinson believed that even false theories play a valuable role in scientific creativity. Such theories were not simply abandoned; rather, they served as raw materials for constructing new, improved theories. Even when conclusively disproved, a theory might direct the scientist's attention to previously unrecognized areas of research. Hutchinson held this view of theory construction throughout his career, but he articulated it most succinctly late in life:

> even a potentially erroneous theory is an enormous advance over having no theory at all, for the incorrectness of the theory, when tested, is in a sense a measure of how far wrong are the postulates on which the theory is based. Once this has been determined, we can start modifying the theory; if we had no theory, there would be nothing to modify and we should get nowhere.[81]

Writing a bit facetiously, Hutchinson once described science as a "dialectical tournament" in which bold new theories entered the lists.[82]

During the clash of ideas, many of these speculations were proven false, but the process of testing led to a deeper understanding of nature. For this reason, he argued that editors ought to be willing to publish speculative, even unorthodox, theoretical articles. On a number of occasions he actively supported younger biologists in their efforts to get controversial ideas into print. For Hutchinson, theoretical ecology was also dialectical in the broader sense that it could be seen in terms of two contrasting approaches: the biogeochemical and biodemographic perspectives. During the late 1940s Hutchinson believed that he was approaching a synthesis of these polarities through the use of formal analogy. Although this ambitious attempt to unify ecology did not succeed, it stimulated an enormous amount of creative research by his students.

Hutchinson's bold approach to theory building shared much in common with that of the young Charles Elton. However, more conservative ecologists were hostile toward what they considered Hutchinson's cavalier approach to research and his theoretical flights of fancy. Just as in his critique of Clementsian theory, Henry Allan Gleason had claimed that every species is a law unto itself, so many limnologists claimed that every lake is a unique individual. General theory, if it could be arrived at, would come only from a painstaking accumulation of facts collected from a wide variety of different aquatic systems. This extreme empiricism was common among limnologists during the pre–World War II era, and it openly conflicted with Hutchinson's approach. Chancey Juday, a prominent limnologist at the University of Wisconsin, sneered:

> The Yale school of mathematical-limnologists is having a high time displaying their mathematical abilities. The interesting part about it is that they are applying mathematical formulae used in sub-atomic physics where all of the forces are presumably uniform to limnological problems where there are all sorts of un-uniform factors involved. . . . Apparently they do not have brains enough to see the point in the two very different situations. . . . In a short time I shall expect them to tell all about a lake thermally and chemically just by sticking one, perhaps two, fingers into the water, then go into a mathematical trance and figure out all of its biological characteristics. As the next stage in their evolution they will probably be able to give a lake an "absent treatment" similar to a spiritualist, so it will not be necessary to visit a lake at all in order to get its complete chemical, physical and biological history.[83]

As we shall see in chapter 5, Juday's antagonism toward Hutchinson's style of research very nearly prevented the publication of one of the most seminal theoretical papers in ecosystem ecology: Raymond Lindeman's "The Trophic-Dynamic Aspect of Ecology."

Intellectual Foundations for a New Ecology

By the end of World War II, the intellectual foundations for ecosystem ecology were in place. Frederic Clements had hammered home the idea that the biological community, a kind of organism, developed ontogenetically. Although much modified during the post–World War II period, both the organismal and developmental aspects of Clementsian ecology became key intellectual components of the ecosystem concept. Closely related to these Clementsian ideas was the suggestion that communities have a form of metabolism. This metabolism consisted of the flow of energy and materials through the community and its surrounding environment. Although not particularly drawn to organismal analogies, Charles Elton pioneered the study of energy flow in his 1927 book *Animal Ecology*. This early attempt at trophic analysis had an important influence on later work. Indeed, energy flow became one of the two most important areas of ecosystem research after World War II. The other was the study of biogeochemical cycling. Hutchinson's influence on this area of research was manifold. He absorbed and expanded the suggestive biogeochemical ideas of V. I. Vernadsky, and he made the writings of the great Soviet scientist available to the American scientific community. Through early studies on Linsley Pond, he and his students played an important role in bringing the new biogeochemistry into ecosystem ecology. Others were doing similar research, but not within a broad theoretical framework. Hutchinson's early forays into cybernetics and mathematical theory provided the rudiments for such a framework. Cybernetics also provided a more sophisticated language for exploring the suggestive similarities between organism and ecosystem.

5

The Birth of a Specialty

Perhaps the most dangerous post for a referee or editor is at the boundary of a well-established discipline where a new subject is striving to be born.

—J. M. ZIMAN, *Public Knowledge: An Essay Concerning the Social Dimension of Science*

CHARLES ELTON AND G. EVELYN HUTCHINSON played crucial roles in establishing the intellectual foundations of ecosystem ecology, but they rarely used the term *ecosystem* in their writings. Hutchinson's work, in particular, seemed to demand a new ecological unit, for he was studying complex interactions not just within the living community but also between the community and its surrounding environment. Writing in 1942, Hutchinson's protégé Raymond L. Lindeman commented that the concept of community forced a "biological emphasis" on a more fundamental set of processes.[1] At what point, Lindeman asked rhetorically, did one draw the boundary between a living community and its nonliving environment? From the perspective of biogeochemistry any such boundary was indistinct, perhaps even arbitrary. Organisms in a lake died, and their protoplasm gradually decomposed into inorganic molecules. At the same time, inorganic materials in the ooze were absorbed by living plants, incorporated into their protoplasm, and transferred through the aquatic food chains. Thus, chemical substances continuously cycled back and forth between the biotic and abiotic worlds. For Lindeman, the ecosystem concept provided a point of departure for exploring this relatively new area of ecological research. It was, he claimed, the fundamental concept in biogeochemistry.

Lindeman's 1942 paper, "The Trophic-Dynamic Aspect of Ecology," is historically significant for a number of reasons. Robert Cook has vividly detailed the difficulties that the young ecologist encountered in his attempts to publish the paper.[2] These difficulties highlight the rapid and fundamental changes that were occurring within the discipline of ecology around World War II. The story is all the more dramatic because Lindeman, then twenty-seven years old, died unexpectedly as the article finally went to press. From the perspective of intellectual history the paper is important because in it Lindeman attempted a synthesis of the ideas of Clements, Elton, and Hutchinson. The result, which outraged some older, more conservative scientists, caught the imagination of a generation of post–World War II ecologists. An effective catalyst, it stimulated the rapid development of ecosystem ecology after the war. The term ecosystem had been used occasionally prior to the 1940s, but Lindeman was perhaps the first ecologist to exploit fully the concept. This chapter explores the origin of the ecosystem concept half a decade before the publication of Lindeman's classic paper, and it analyzes the way that the young ecologist used this idea to establish a new ecological specialty.

Arthur Tansley and the Ecosystem Concept

Ecosystem was coined in the mid-1930s by the British plant ecologist Arthur Tansley (figure 6).[3] Because it first appeared in a strongly worded critique of Clementsian ecology, the concept has been interpreted as a radical break from Frederic Clements's idea of the organismal community. Donald Worster, in *Nature's Economy,* goes so far as to claim that Tansley's ecosystem concept owed nothing to earlier biological ideas.[4] And Ronald Tobey, in his detailed history of American grassland ecology, characterizes Tansley's critique as a "vigorous, nearly bitter, attack" on Clementsian ecology.[5] According to Tobey, the publication of Tansley's 1935 essay signaled the breakup of a Clementsian "microparadigm."[6] For Tobey, Tansley serves as a kind of intellectual barometer, an indicator of the scientific acceptance of Clements's ideas. According to this interpretation, Tansley was an early supporter of Clements, but this support steadily diminished and culminated in his public rejection of Clementsianism in 1935. I share Tobey's belief in the historical importance of the relationship between Tansley and Clements; however, my interpretation of this episode is somewhat different, primarily because I am unwilling to accept

Tobey's Kuhnian perspective on history. By presenting the history of ecology as the rise and fall of paradigms, this perspective places too much emphasis upon historical discontinuity. In chapters 2 and 3, I argued that the reception of Clements's ideas was mixed. Many leading ecologists accepted some of Clements's ideas, but rejected others; they did not accept a complete "Clementsian paradigm." In my view, Tansley exemplified this type of critical attitude. From the beginning, Tansley's enthusiasm for Clementsian ecology was tempered by considerable skepticism. And although his 1935 paper was particularly outspoken, it was not uncharacteristic from a scientist known for biting, incisive analysis. In contrast to Tobey's conclusion that this article represented a rejection of a Clementsian paradigm, my interpretation is closer to that suggested by one of Tansley's students, his biographer Harry Godwin. According to Godwin, when Tansley put forward his ecosystem concept, he was "qualifying without disabling" Clements's earlier organismal concept.[7] My claim, in contrast to the conclusions of both Worster and Tobey, is that after World War II the ecosystem concept continued to reflect important elements of Clements's thinking. Tansley's sophisticated mechanistic view of nature also retained strong organismal overtones.

Tansley and Clements carried on a warm correspondence over the course of several decades beginning in 1905.[8] Aside from a common interest in ecology, there was little similarity between the urbane Tansley and his more parochial American friend. Born in 1871, three years before Clements, Tansley was the product of upper middle-class Victorian society.[9] In contrast to Clements whose serious interests were limited to botany, Tansley was a self-described dilettante. As a student at Cambridge he was a friend of Bertrand Russell, and he maintained an avocational interest in philosophy throughout his career. "If you scratch a biologist," a colleague later remarked of Tansley, "you will find a philosopher."[10] As a young man he assisted the aging Herbert Spencer in revising his *Principles of Biology.* During World War I he became interested in psychoanalysis and wrote a popular book on Freudian psychology and its relationship to modern biology. This interlude in his botanical career culminated when he moved to Vienna for several months to study under Freud.

Tansley's scientific work lacked the brilliant originality of Clements's studies, but his contributions to professional ecology were nonetheless profound. As editor of *The New Phytologist,* a journal he started with his own money in 1902, Tansley promoted ecology and other new areas of botanical research. He was instrumental in establishing the British Ecological Society.[11] When the group was formed in 1913,

Tansley was elected president, a post that he held again during the late 1930s. For twenty years he served as editor of the society's *Journal of Ecology*. When he retired in 1937 Tansley was the most influential ecologist in Britain. This influence rested as much upon his skills as a scientific leader as on the results of his research.

Tansley's editorial duties undoubtedly sharpened his prodigious abilities as an insightful critic. According to Harry Godwin, Tansley had a unique ability to grasp new ideas and see their potential usefulness.[12] I believe that this talent, combined with his broad intellectual interests, and his leadership position within the rapidly growing discipline explain Tansley's attitude toward Clementsian ecology. He was sympathetic toward Clements's claim that a community is a kind of organism, an idea that he considered inherently useful, but he was too sophisticated to accept it completely. In a long series of critical review articles Tansley defended Clements's organismal ideas, but he also reworked these ideas into a philosophically more acceptable form. The ultimate result of this process was the ecosystem concept.

Tansley's initial foray into Clementsian ecology came in a lengthy review of Clements's first major book, *Research Methods in Ecology* (1905). The review, which appeared in his newly established journal, was written with Tansley's brother-in-law, the Cambridge physiologist F. F. Blackman. Although Clements later thanked Tansley for his "generous praise" of the book, the review had, in fact, contained much serious criticism.[13] Blackman and Tansley considered *Research Methods* the most ambitious and important recent contribution to ecology. They were particularly impressed by Clements's discussion of ecological succession, and they were generally sympathetic to his organismal concept of the plant community: "His view of vegetation as an organism is as legitimate as the familiar idea of a human society from the same point of view. Both conceptions are useful and desirable so long as it is remembered that they are essentially analogical, that these quasi-organisms do not possess many of the essential features of real organisms."[14] Clements had based his organismal theorizing on a belief that ecology was a branch of physiology: "a rational field physiology."[15] His reviewers were decidedly unsympathetic toward such physiological pretensions. Indeed, Blackman nearly refused to coauthor the review because he was so "repelled" by Clements's discussion of physiology. In a letter to Tansley, Blackman complained that the entire section on physiological principles was "dreadfully crude" and "glib."[16] Coming from a close friend, a scientist he deeply admired, Tansley must have taken this private dismissal of Clements's ideas seriously. However much he was attracted to

Clementsian ecology, Tansley always maintained some intellectual distance from the American ecologist. The published review of Clements's first book, a mixture of profound admiration and sharp criticism, perfectly encapsulated Tansley's attitude toward Clementsian ecology, an attitude that found expression in several articles he wrote during the next thirty years.[17]

Tansley was no stranger to controversy; as editor he freely expressed his opinions on a range of scientific topics. A reader, stung by one of his editorials, once accused Tansley of fomenting "botanical Bolshevism" in the pages of the *New Phytologist.*[18] The comment appears ironic given Tansley's later opposition to Marxism, but it illustrates Tansley's willingness publicly to defend or criticize ideas, regardless of their popularity. During the twenty years following the publication of *Research Methods in Ecology,* Tansley promoted Clementsian ecology and he vigorously defended the organismal concept against vociferous European critics. But this does not mean that he became a disciple of Clements. In their correspondence the two men frankly acknowledged their theoretical differences, and in his public statements Tansley was equally blunt in criticizing certain aspects of Clementsian ecology. From Tansley's point of view, his American friend was arguing an extreme position, and he frequently made inexcusable philosophical errors.

To understand the origin of the ecosystem concept, it is important to consider in greater detail Tansley's critical attitude toward Clementsian ecology. Nature might be composed of aggregations of individuals, as Henry Allan Gleason argued, but Tansley responded that focusing only upon the individuals led the ecologist to ignore important biological processes.[19] The ecologist had to consider the forest as well as the trees, and the idea of an organismal community drew attention to interactive processes such as succession. By emphasizing process, Clements was shifting ecology away from mere description. This, Tansley believed, was tremendously useful, for, if nothing else, it stimulated the intellectual development of the young discipline. This support for the organismal community concept was tempered, however, by Tansley's critical attitude and his philosophical training. "I . . . believe the analogy with the organism to be legitimate and useful," the British ecologist wrote, "if it is not pushed too far, and especially if we abstain from making illegitimate deductions."[20] The problem was that Clements did push the analogy too far and did make illegitimate deductions. Most egregiously, he often confused analogy and true identity. For Clements, succession was not just analogous to ontogeny; it was ontogeny. Because succession was a form of development it had to be progressive; the community always matured toward a well-defined climax. Like most ecologists, Tansley rejected this claim. In some cases, which he termed *autogenic*, succession could be thought of as a developmental process. As plants successively modified the environment, the community slowly changed in an orderly manner, ultimately reaching the mature climax state. In cases of *allogenic* succession, however, fire or some other environmental factor so disrupted the process that several successional stages might occur simultaneously in the same area. Clements was quite familiar with this phenomenon, but he often ignored it in his theoretical writings. According to Tansley, one could hardly see "development" in a situation where nature was constantly in a disturbed state.

To avoid Clements's logical errors, Tansley suggested that ecologists refer to plant communities not as organisms, but as quasi-organisms. However, even this construction led to misunderstanding, and toward the end of his career Tansley proposed the more neutral term *ecosystem*. He introduced this new term in his most detailed critique of Clementsian ecology: "The Use and Abuse of Vegetational Concepts and Terms." This critical essay, which appeared in 1935, was aimed not at Clements directly, but rather at three review articles written by the South African ecologist John Phillips.[21] Although Phillips presented a comprehensive discussion of the current state of community ecology, he made no attempt to hide his partiality to Clements's ideas. At the conclusion of the final article, Phillips "earnestly" invited criticism from his readers.[22] For Tansley, with his keen analytical mind and critical temperament, Phillips's request amounted to throwing down the gauntlet.

Clements proudly claimed that Phillips's articles presented the organismal concept to a new generation of ecologists.[23] But from Tansley's point of view, the young South African exemplified the worst features of Clementsian ecology. In his articles, Phillips blatantly stepped over the line that separated useful analogy from dangerously misleading ideas. According to Tansley, "Phillips' articles remind one irresistibly of the exposition of a creed—of a closed system of religious or philosophical dogma."[24] Hyperbole notwithstanding, there was much in Phillips's papers that offended Tansley's basic scientific beliefs. For Tansley, the scientist had to be pragmatic, always willing to modify or reject concepts when they were found wanting, but Phillips fairly bubbled over with uncritical enthusiasm for holistic organicism, claiming that "it has become to me the deepest and most abiding reality, paradoxically both a starting point and a goal in the scientific study of communities."[25] With the same uncritical zeal, Phillips derived illegitimate deductions from the organismal concept, restating Clements's threadbare claim that succession was always progressive. This must have been particularly irritating to Tansley, given the distinction between autogenic and allogenic succession that he had made so carefully a decade earlier. Rather than providing a detailed refutation of Tansley's distinction, however, Phillips blithely stated that Clements and his followers refused to recognize the existence of allogenic succession. Given such a transparently partisan position, there was some justification for Tansley's acerbic claim that Phillips's entire discussion of succession rested upon an article of faith: the community is an organism; therefore, it *must* develop to a single, adult form.

Tansley was also critical of Phillips's organicism on more general philosophical grounds. In the final article of his series, Phillips allied his organismal ideas with the philosophy of biology presented in Jan Smuts's inscrutable *Holism and Evolution.*[26] This was a significant change in Clementsian ecology, for in his early writings Clements had rarely explored the broader philosophical implications of his organismal ecology. Prior to the 1930s he apparently was unfamiliar with the holistic organicism of biologists such as John Scott Haldane and William Morton Wheeler. And on the face of it, Clements's mechanical-organic theory of succession fit uncomfortably with Smuts's passionately antimechanistic defense of emergent evolution. Furthermore, Smuts specifically denied that the organismal community, at least in human sociology, was anything more than a figure of speech.[27] Nonetheless, by 1935 Clements was avidly recommending Smuts's book to his colleagues, and both he and Phillips were using holism as a philosophical justification for their organismal ecology.[28]

Tansley objected to holism on almost every point, objections shared by several other critics of Smuts's book.[29] For Tansley, Smuts's claim that the properties of a whole cannot be predicted from those of its parts amounted to intellectual defeatism. An example of such emergent properties, one used by Smuts and Phillips, was that the properties of a water molecule cannot be predicted solely from the properties of hydrogen and oxygen atoms. To this, Tansley retorted, "who will be so bold as to say that this new entity, for example the molecule of water and its qualities, would be unpredictable, if we really understood *all* the properties of hydrogen and oxygen atoms and the forces brought into play by their union? Unpredictable by us with our present knowledge, yes; but *theoretically* unpredictable, surely not."[30] To dogmatically assert that reductionistic explanations were invalid amounted to slamming the door on a potentially fruitful approach to scientific research. Smuts's claim that the whole was more than the sum of parts was also objectionable to Tansley because the term *sum* was so vague. Indeed, Smuts's entire discussion of holism and its application in the sciences was marked by a singular lack of clarity. Given his penchant for logical analysis, Tansley must have found deciphering the meaning of holism a trying experience. Too often Smuts's writing degenerated into numbingly repetitive jargon: "The whole-making, holistic tendency, or Holism, operating in and through particular wholes, is seen at all stages of existence."[31] The claim that this nebulous holism was a "creative force" was dangerously misleading. In a certain sense, Tansley admitted, a community of plants (a "whole") might be said to be the cause of its own development, but this really meant nothing more than that under certain circumstances succession resulted from the collective activities of a set of individual plants. There was no "holistic tendency" at work in succession, only specifiable interactions among individuals and between these individuals and the physical environment.

The intensity of Tansley's attack on Phillips's holism may have reflected deeply held political and social beliefs, a historical interpretation suggested by Ronald Tobey.[32] An admirer of Herbert Spencer, Tansley was deeply committed to individualism.[33] In his early writings, Clements rarely explored political or philosophical issues. However, during the 1930s, he began to espouse a collectivist view of society. Decrying the "myopic individualism" that he considered responsible for the tragedy of the Dust Bowl, Clements applauded the social engineering of the New Deal.[34] Society was a huge organism greater than the sum of its parts; its efficient operation required extensive social integration directed by an elite group of technical experts. Within the intellectual context of the 1930s, there was nothing particularly original or profound in Clements's social views,[35] but they touched a raw nerve in Tansley. Collectivism was abhorrent to him; beneath its biological veneer, the holism espoused by Smuts, Clements, and Phillips embodied an objectionable political philosophy.[36]

Tansley's students have vigorously denied that his scientific views were shaped by political or philosophical beliefs. It is certainly true that Tansley's ecological ideas developed many years before his involvement in the Society for Freedom in Science and his public statements against Marxist totalitarianism during World War II.[37] Furthermore, when it suited his philosophical purposes, Tansley used the ideas of prominent Marxist thinkers. Tobey's interpretation, however, cannot be rejected out of hand. Tansley's writings reveal a man who refused to distinguish sharply between biology and political philosophy. More than most ecologists of his day, Tansley was willing to draw analogies between nature and human society. There may have been no direct cause-and-effect relationship, but it seems reasonable that the ecosystem concept developed partly as a result of Tansley's impatience with the social implications of holistic organicism.

Despite his devastating critique of Phillips's ideas, Tansley never completely broke with the organismal concept. As he had several times before, Tansley restated his belief that important analogies existed between organisms and communities. The problem, as he had repeated so often, was that Clements (and now Phillips) pushed the analogies too far. They continued to insist that communities were not simply like organisms but that they were, in fact, organisms. This intransigence was impeding the growth of ecology. Tansley, therefore, proposed the ecosystem as an alternative concept that would avoid the excesses of Clementsian ecology without sacrificing the inherent usefulness of the organismal concept. As Tansley later acknowledged, one could still consider the ecosystem to be like an organism, but such a belief was not necessary.[38] Indeed, Tansley seemed to prefer a concept with a pronounced flavor of modern physics.

Tansley based his new ecosystem concept on the philosophy of Hyman Levy, a prominent British Marxist and mathematical engineer. According to Levy, the universe is composed of partially overlapping systems. "Science," Levy argued, "like common sense, sets out in the first instance to search for systems that can be imagined as isolated from their setting in the universe."[39] These mental "isolates," which always corresponded to some physical reality, could be expanded or reduced according to purposes of the scientific analysis. For example, for the purposes of anatomy, a tree could be considered a well-isolated biological system, but for the physiological purposes of understanding photosynthesis or growth, the system needed to expand to also include some elements of the physical environment.[40]

Levy's *The Universe of Science* presented a picture of science similar to the more fragmentary one that Tansley had accepted throughout the 1920s and 1930s. Because nature was so complex, ecologists were forced to focus on only certain aspects of "kaleidoscopic" reality.[41] In an important sense, communities and ecosystems were natural objects, for they corresponded to something real in nature; but they were also abstractions, isolated from the welter of sense experience. "Actually the systems we isolate mentally are not only included as parts of larger ones," Tansley argued, "but they also overlap, interlock and interact with one another. The isolation is partly artificial, but is the only possible way in which we can proceed."[42] This methodological statement was tremendously significant for the future of ecology. It freed the ecologist from the rigid organismal concept advocated by Clements. One could still draw useful analogies between organisms and ecosystems, but there was no reason to believe that the ecosystem really was a type of organism. Tansley's innovation also freed the ecosystem concept from a rigid geographical basis. The concept of community, as a geographically defined entity, had become mired in controversy. What were the natural boundaries of communities? Were communities local associations of plants, or did they extend over climatic regions? These controversies had been incessant and inconclusive, but the ecosystem concept made them irrelevant. Ecosystems were not completely arbitrary; for example, a lake might have fairly well-defined boundaries. Even in situations where obvious boundaries did not exist, however, the ecologist could define them for the purposes of research.

Tansley was not an experimental ecologist, and apparently he was only dimly aware of the possibilities implied by his new concept. But G. Evelyn Hutchinson recognized the necessity of isolating systems, perhaps even arbitrarily, to measure the transfer of energy and matter across their boundaries. This experimental approach was used with great success in fields such as thermodynamics, biochemistry, and physiology, and by the end of the 1930s Hutchinson saw that it could also become the starting point for a new ecology.[43] Within half a decade, Raymond Lindeman set the course for this new ecology.

Raymond Lindeman and a New Direction for Ecology

Raymond Lindeman was an earnest young man (figure 7). "I don't believe I ever saw him laugh or smile," recalled a friend. "He was subject to stomach ulcers. . . . He was so intense with his field studies that sometimes he would work until he began to vomit blood."[44] That recollection may be overly severe, but by all accounts Lindeman became almost totally immersed in his research. As a high school student applying to Park College in Missouri, he expressed a desire to become an experimental biologist. He had considered law as a more practical alternative, he wrote, but had given up this "passing fancy" for his first ambition—science. Despite the financial attraction of law, a life in science provided a greater opportunity to serve humanity, and it allowed one to "see and try to understand the majestic symmetry of the universe."[45] This was not simply an exaggerated expression of youthful idealism. Growing up during the Great Depression, the son of a struggling Minnesota farmer, Lindeman knew the value of a dollar. Rather, this early statement spoke to a deeper idealism and commitment to science, personal characteristics that made a deep impression on those who knew him.[46]

After graduation from Park College in 1936, Lindeman studied limnology under Samuel Eddy at the University of Minnesota. Eddy was a competent, but not particularly distinguished, scientist. Of greater importance to Lindeman's intellectual development was the plant ecologist W. S. Cooper, a student of Henry Chandler Cowles. In his plant ecology courses, Cooper presented a broad historical overview of the field. The leading figures in the field, from the late nineteenth century on, were discussed and their ideas criticized. Lindeman took copious notes during Cooper's lectures and later highlighted the important points with colored pencil.[47] These theoretical issues were also discussed in a more informal setting when graduate students met regularly at the Cooper home.[48] Thus, by the time that he finished his graduate studies, Lindeman was not only trained as a limnologist, but, under Cooper's tutelage, he had also been exposed to the major conceptual developments shaping pre–World War II plant ecology. These ideas, particularly the concepts of succession and climax community, held an important place in Lindeman's later theoretical writings.

Lindeman came to Hutchinson's laboratory in September 1941 on a Sterling postdoctoral fellowship. The fellowship year, shortened by his death in June 1942, was a remarkably creative period for Lindeman. His brief association with Hutchinson resulted in a classic paper, "The Trophic-Dynamic Aspect of Ecology." In this paper Lindeman used the new ecosystem concept to create a synthesis of ecological principles derived primarily from Clements, Elton, and Hutchinson. The paper formed the cornerstone for much post–World War II ecosystem ecology. Initially, however, Lindeman's theoretical approach met with considerable skepticism and hostility from several leading ecologists. The manuscript was originally rejected by the editorial board of *Ecology*, and only through Hutchinson's intervention was it finally accepted by that journal.[49] The story is all the more poignant because Lindeman died from a congenital liver ailment shortly before his paper appeared in print.

Great research and great teaching do not necessarily go together, but in the case of Hutchinson the two activities meshed perfectly. One key to his success was his ability to attract imaginative young biologists to his laboratory and to stimulate their intellectual development. What often resulted from these collaborations were highly original theoretical works woven together from ideas provided by both teacher and student. No case illustrates this form of collaboration so clearly as Lindeman's article on trophic dynamics.[50] In this paper Lindeman credited Hutchinson with providing many key concepts that he used in his theoretical discussion of limnology. A superficial reading of the paper might give the impression that the truly original parts of the paper were Hutchinson's rather than Lindeman's. But archival evidence suggests a very different interpretation. Writing the paper involved a complex development of ideas, ideas that neither man had fully worked out before Lindeman arrived at Yale.

One can trace the development of these ideas by comparing Lindeman's dissertation and the successive drafts of the trophic dynamic paper.[51] Analyzing Hutchinson's influence on this development is complicated by the fact that Lindeman was quite familiar with Hutchinson's work when he was writing his dissertation in Minnesota. Throughout this period (1939–1941) Lindeman corresponded frequently with Hutchinson's student Edward Deevey and somewhat less frequently with Hutchinson himself. This correspondence reveals that Lindeman began thinking deeply about Hutchinson's approach to ecology well before he arrived in New Haven. If Lindeman wrote anything during his "pre-Hutchinsonian" period, it apparently no longer exists.

Less directly, the historian can ask what each man brought to the relationship. By 1941, Hutchinson had spent a decade studying the biogeochemistry of Linsley Pond. He was beginning to establish his reputation as one of the leading limnologists in the United States, and he was sufficiently confident of his position within the specialty that he had begun writing a general treatise on limnology.[52] He had also taken very preliminary steps toward outlining a mathematical explanation for energy flow. Each trophic level could be characterized by its energy content, which changed as energy entered and left. Therefore, a complete explanation of community energetics consisted of a series of differential equations representing movements of energy into and out of the various trophic levels. His sketchy notes formed a suggestive beginning, but as Hutchinson admitted, it was hardly a theory: "At present it is merely possible, introducing certain assumptions of a more or less arbitrary nature, to obtain approximate values in the very simplified equations that result."[53]

There is no reason to believe that Lindeman contributed directly to the development of Hutchinson's mathematical concepts. His background in mathematics amounted to no more than courses in trigonometry and college algebra and perhaps a graduate course in statistics.[54] Indirectly, he undoubtedly contributed a great deal, for he brought with him a wealth of information about trophic relationships in aquatic systems. As a graduate student Lindeman had spent five years studying a senescent lake in central Minnesota: Cedar Lake Bog. His dissertation dealt broadly with the geological, physical, chemical, and biological characteristics of the lake.[55] Most of the work focused, however, upon the food web in the lake. Prior to 1941 Hutchinson had not studied the trophic relationships in Linsley Pond in any detail. Thus Lindeman provided him with a fund of knowledge and field experience necessary for building a mathematical theory. Lindeman, in turn, used Hutchinson's nascent mathematical concepts to develop a very general theory of energy flow in ecosystems. He had begun to explore these ideas in a tentative and qualitative manner in the final chapter of his dissertation. During his brief stay at Yale, he found the mathematical tools to do so in a much more rigorous fashion.

In the opening paragraphs of his article, Lindeman presented the "trophic-dynamic viewpoint" as a new, more fundamental approach to ecological problems. Earlier ecologists, according to Lindeman, had emphasized the taxonomic composition of communities; they had described either the static distribution of species or the changes in species during the course of succession. Repeating a criticism that Hutchinson had previously voiced, Lindeman argued that these earlier approaches—what he referred to as the "static species distributional viewpoint" and the "dynamic species distributional viewpoint"—drew an unnatural dividing line between the living community and the nonliving environment.[56] Biogeochemistry was obliterating this arbitrary distinction; energy and materials were constantly moving back and forth across the often indistinct boundary separating the organic and inorganic worlds. "This constant organic-inorganic cycle of nutritive substance is so completely integrated," Lindeman argued, "that to consider even such a unit as a lake primarily as a biotic community appears to force a 'biological' emphasis upon a more basic functional organization."[57] This more basic functional organization was found at the level of the ecosystem, a "system composed of physical-chemical-biological processes active within a space-time unit of any magnitude."[58] Taking Tansley's skeletal concept, Lindeman fleshed it out. He took Charles Elton's older trophic dynamic theory, which had focused almost entirely upon the living community, and, using the newer concept of ecosystem, he gave this theory a biogeochemical interpretation.

Elton's early trophic theory had traced the movement of *food* through the living community, and in his earlier writings Lindeman accepted this general explanatory scheme. After arriving at Yale and reading Hutchinson's lecture notes, Lindeman began to distinguish more clearly between *energy* and *matter*. This important conceptual refinement is clearly revealed in the food cycle diagrams that Lindeman included in his early publications.[59] Chemical substances underwent constant cycles of synthesis and degradation. Simple inorganic compounds were synthesized into complex organic molecules by photosynthetic organisms, the *producers* of the ecosystem. These organic molecules underwent further chemical transformations as they moved through food chains composed of herbivores and carnivores, the *consumers* of the ecosystem. Ultimately, all these complex organic molecules were degraded again to simple inorganic substances by the action of bacteria and other *decomposers* in the ecosystem.

Driving this cyclic process was a one-way flow of energy. Solar energy was captured during photosynthesis and transformed into chemical energy stored within organic molecules. Through the process of predation this chemical energy was transferred from one trophic level to the next, but at each step some was converted into kinetic energy or otherwise dissipated to the environment. Thus, not all chemical energy stored in a trophic level could be transferred to the next higher level. This was of fundamental importance because it meant that, unlike matter that could be recycled through the ecosystem, a constant flux of energy was necessary to keep the ecosystem operating. This realization also suggested a more fundamental solution to an important biological question that Charles Elton had tentatively answered a decade and a half earlier: Why is the number of trophic levels in a food chain generally limited to four or five?

Elton had answered this intriguing question in terms of the size and numbers of organisms. As one moved up the food chain, animals tended to become larger and fewer in number. At some point, generally around the fourth or fifth step in the chain, the surplus food produced by a trophic level was no longer able to support another group of even larger predators. Lindeman solved this problem more elegantly by reducing it to a physical explanation in terms of energy: "the energy of no food level can be completely extracted by the organisms that feed upon it."[60] Formally, one could represent this relationship among trophic levels by a series of terms:

$$\lambda_0 > \lambda_1 > \lambda_2 \ldots > \lambda_n$$

where the productivity, λ_n, represented the energy transferred to a trophic level from the next lower level. As one moved up the food chain, productivity decreased at each step. Thus, Lindeman had reformulated Elton's pyramid of numbers into a pyramid of energy. Available energy decreased at each successive level in a food chain. At some point, generally the fourth or fifth level in the chain, the amount of energy was insufficient to sustain another population of predators.

Why did productivity decrease at each successive trophic level? Each trophic level (n) contained a quantity of energy (Λ_n). Energy entered the trophic level at a rate (λ_n), which Lindeman had defined as the productivity of level n. At the same time energy was lost at a rate (λ_n'). But this energy loss from trophic level n was not the same as the energy that entered the next trophic level (Λ_{n+1}). In fact, it was always greater. Much energy was dissipated from trophic level n as a result of metabolism. Although this respiratory heat constituted a loss of energy from the quantity Λ_n, it did not constitute a gain in Λ_{n+1}. Furthermore, some organic matter consumed by the predators in level $n + 1$ was not assimilated. Wood, hair, bone, and other relatively indigestible substances contain calories, but not in a form usable by most animals. Again, these calories constituted a loss from Λ_n, but they did not contribute to Λ_{n+1}. Finally, some members of trophic level n died of nonpredatory causes. These calories were transferred immediately to decomposers, not to the consumers at trophic level $n + 1$.

Although Lindeman did not use such pictorial representations in his paper, one can visualize the energy budget for trophic level n as a simple input-ouput system, with the energy input always exceeding the useable output to the next higher level. Each trophic level is a "black box" containing a quantity of energy (Λ_n). Energy flows into the box at the rate (λ_n) and flows out of the box at a rate (λ_n'). Not all this energy flows, however, into the box representing the next trophic level; λ_n' is always greater than λ_{n+1}. The difference represents either energy lost as heat or energy in a form unusable by organisms in trophic level $n + 1$. This type of black box diagram became a standard method of representing trophic dynamics during the decades following World War II (figure 8).

Lindeman's formulation posed a number of intriguing questions. For example, how do the productivities of the same trophic level in different ecosystems compare? And within the same ecosystem, is the efficiency of energy transfer between any two successive trophic levels (i.e., λ_{n-1}/λ_n) a constant, or does it vary? Providing answers to such questions was central to understanding both how ecosystems function and perhaps how to exploit them economically. Elton had suggested that all ecosystems share a common trophic "ground plan"; Lindeman's paper suggested how this ground plan might be quantitatively studied. For young ecologists entering the field after World War II this was an exciting prospect.

The final section of Lindeman's paper was the most speculative, for here he attempted to unite succession, a central concept in pre–World War II ecology, with the newer study of energy flow. His discussion of succession was heavily influenced by both older Clementsian notions of development and Hutchinson's developmental metaphors for productivity in Linsley Pond. A few years before meeting Lindeman, Hutchinson had suggested that as a lake aged productivity followed a logistic pattern of growth, reaching what he referred to as "trophic equilibrium" when the lake became nutrient-rich, or eutrophic. In the original draft of his paper, Lindeman expanded on this Hutchinsonian idea by portraying succession as a series of such sigmoid curves. Each stage in succession, from lake to bog to forest, was accompanied by an increase in productivity. Productivity increased exponentially at first, but then plateaued as the system reached a new "stage equilibrium." This successive pattern of logistic growth continued until a climax forest was established and productivity was maximized. Lindeman and Hutchinson presented this successional scheme at the annual meeting of the Ecological Society of America in Dallas, and the younger ecologist later acknowledged that it came as "a rude shock" when he was informed that there was no empirical support for his simple growth curves for productivity.[61] After some tinkering, he settled on a less elegant graph. In the published version, one can still see the remnants of the original sigmoid curves. But the transition from aquatic to terrestrial succession is marked by a dramatic decrease in productivity, and productivity also decreases slightly at climax.

As Lindeman readily admitted, his attempted synthesis of succession and energetics, based on meager data, was necessarily tentative; however, it foreshadowed a number of more sophisticated discussions of the energetic basis of succession.[62] And the evolution of this section of the paper is intriguing for the glimpse it provides of Lindeman's creative imagination. His first impulse was to envision succession as a progressive, developmental process leading to a climax state. This reflected the modified Clementsian ideas that he had been taught by W. S. Cooper at the University of Minnesota.[63] It also reflected Hutchinson's view that succession was a developmental process leading to a stable equilibrium. Following Hutchinson's lead, Lindeman idealized this process by using a simple mathematical formula, the logistic equation.

But Lindeman was neither a dogmatic Clementsian nor a detached theoretician. When he realized that his original explanation could not be supported by empirical evidence, he was willing to abandon it—but not completely. In true Hutchinsonian fashion, he used elements of the falsified scheme to produce a more realistic picture of succession.

Lindeman and His Critics

Ecosystems, productivities, and trophic levels have become so embedded in the language of modern ecology that it is difficult for the modern reader to recognize just how revolutionary Lindeman's paper appeared in 1942. It stands as one of the great intellectual watersheds in the history of ecology. On one side of this divide stood a small group of mostly younger ecologists who rather quickly grasped the significance of Lindeman's accomplishment. The publication of his paper was followed by a flurry of similar attempts to refine, clarify, and expand trophic concepts.[64] These, together with important field studies, were largely responsible for the early development of ecosystem ecology after World War II. On the other side of this intellectual watershed stood a larger group of ecologists, older and well-established in the field, who had little appreciation of Lindeman's accomplishment.

Two months after arriving in Hutchinson's laboratory, Lindeman submitted his paper for publication in the journal *Ecology*.[65] The responses from two anonymous reviewers, now believed to have been Chancey Juday and Paul Welch, were extremely negative.[66] Welch, a professor at the University of Michigan and author of a highly respected textbook on limnology, complained that the paper was too speculative and lacking in empirical evidence; in short, it was a "desk produced" essay, rather than a true piece of scientific research. "What limnology needs now most of all," Welch concluded, "is research of the type which yields actual significant data rather than postulations and theoretical treatments."[67] Juday, a biologist at the University of Wisconsin and perhaps the leading limnologist in the United States, was even more condemnatory. As discussed in chapter 4, Juday was bitterly opposed to the mathematical ecology of Hutchinson and his students. His commentary on Lindeman's paper dripped with sarcasm: "A large percentage of the following discussion and argument is based on 'belief, probability, possibility, assumption, and imaginary lakes' rather than on *actual* observation and data."[68] "According to our experiences," Juday continued, "lakes are *'rank individualists'* and are *very stubborn* about fitting into mathematical formulae and artificial schemes proposed by man."

Thomas Park, the editor of *Ecology,* was placed in a quandary. He found Lindeman's paper stimulating but was reluctant to publish the manuscript over the strident objections of two eminent limnologists. With "some reluctance and distress," Park wrote Lindeman that he was forced to reject the paper.[69] Almost immediately, Hutchinson, in a three-page letter to Park, defended Lindeman's theoretical approach and rebutted the referees' criticisms. Park sent copies of this letter to Juday and Welch, who, after reading it, restated their strong opposition to publishing Lindeman's paper. Faced with this impasse, Park allowed Lindeman to resubmit the paper, which he then turned over to his colleague at the University of Chicago, Warder Clyde Allee. Allee, unenthusiastic about the paper, was not strongly opposed to having it published. Despite this lukewarm response from a third referee, Park decided to accept the paper. "I rather imagine that the original referees will still object to certain of its basic premises," he wrote Lindeman, "but I think it best to publish your paper regardless. Time is a greater sifter in these matters and it alone will judge the question."[70]

During the negotiations over his paper, Lindeman suffered an attack of jaundice, a disease he also had suffered from as a graduate student. His health deteriorated rapidly during the spring of 1942 and following surgery he died on June 15. Four months later, his trophic-dynamic paper appeared in *Ecology.*

The initial negative reactions to Lindeman's paper reflect resistance to fundamental changes occurring in ecology, changes that closely accompanied the birth of the new specialty of ecosystem ecology. As Robert Cook has pointed out, much antipathy toward Lindeman's work stemmed from a strong mistrust of mathematical theory.[71] This aversion to mathematics is clearly evident both in Juday's review of Lindeman's article and his other vitriolic attacks on the Hutchinson school.[72] But more was at issue here than mathematics. Juday was a leader of the most prestigious limnological research group in the United States. Philosophically he was a strong empiricist, leery of all general theories—mathematical or not. The Hutchinson school was much smaller, but it was rapidly gaining influence. Both the philosophical differences and the rivalry between the two groups is evident in the sarcastic jokes that circulated privately in the two competing laboratories. Juday derided the Hutchinson team for analyzing lakes by dipping a finger or two into the water.[73] Hutchinson's students condescendingly remarked that Juday's team was capable of making ten thousand measurements without having a single idea.[74] Inevitably, the two groups viewed Lindeman's paper differently.

From the perspective of Juday and Welch, Lindeman's paper was speculative because it attempted to draw broad generalizations from a rather limited body of data. Lindeman had relied primarily upon his own detailed study of Cedar Lake Bog, but he also referred to several other limnological studies, including one by Juday. This was not enough for the Wisconsin limnologist. Lakes were "rank individualists," Juday claimed, resistant to theoretical generalizations. Understanding limnology could only come from the gradual accumulation of data gathered from many different lakes, a painstaking empirical process that Juday had begun several years before Lindeman was born. Other biologists agreed. Although he wrote a sympathetic letter to Hutchinson after Lindeman's death and expressed enthusiasm for the general thrust of the young scientist's research, Charles Elton later complained that Lindeman's calculations were "guesses rather than reliable facts. . . . His conclusions are, to say the least, a shot in the dark."[75]

In marked contrast to Juday's "bottom-up" approach to ecological research, Hutchinson and Lindeman saw a much more creative role for theory. Linsley Pond and Cedar Lake Bog were model systems from which general explanatory principles could be derived. These principles could be empirically tested, but even if they eventually turned out to be false they served as useful guides to further research. Both men believed that this justified publishing the trophic-dynamic paper. In a letter to W. S. Cooper, Lindeman admitted that his theoretical work had a rather shaky empirical base. "I have a feeling, though," he wrote, "that at least some of the ideas are piquing enough to start some people making ecological studies on the basis of productivity and efficiency, and that would be quite gratifying even though some of the hesitantly proposed 'principles' turn out to be wrong."[76] "Even should none of his [Lindeman's] generalizations ultimately hold," Hutchinson wrote Park after the manuscript's rejection, "the work of disproving them will provide important information that would probably be obtained in no other way."[77] From the perspective of Juday and Welch, Lindeman's desk-produced paper was taking up space in a journal that ought to be devoted to hard data. But from the perspective of Lindeman and Hutchinson, the speculative paper served as both guide and stimulus for future research.

With historical hindsight it is easy to see Lindeman's paper as the major point of departure for post–World War II ecosystem ecology,[78] but both Welch and Juday failed to recognize Lindeman's role in establishing a new specialty. Their failure to do so is perhaps not too surprising. Juday had constructed energy budgets for Wisconsin lakes, but he had never attempted to place these within a broader theoretical framework. Others, notably Edgar Transeau at Ohio State University, had taken a more theoretical approach to ecological energetics. However, Transeau had limited his discussion to the accumulation of energy by plants, what is now referred to as primary productivity.[79] Charles Elton and the German limnologist August Thienemann had independently introduced the concept of trophic levels several years earlier.[80] But what most clearly distinguished Lindeman from older ecologists was his use of the ecosystem concept. With this concept he was able to synthesize elements taken from traditional studies in fisheries biology and limnology, the newer biogeochemical approach to studying lakes, Elton's terrestrial animal ecology, and traditional plant ecology.

Lindeman himself believed that he was working within the "Eltonian tradition."[81] But, as Elton's lukewarm reaction to the later development ecosystem ecology suggested, Lindeman's approach differed significantly from the British ecologist's "scientific natural history." The ecosystem concept was better suited to the type of highly abstract theorizing that Lindeman and Hutchinson employed. Lakes, streams, forests, or coral reefs could all be treated as idealized systems composed of trophic levels. Each trophic level was a kind of black box through which energy and materials moved. Indeed, as Hutchinson noted in a postscript to the trophic-dynamic paper, Lindeman had reduced all the biological complexity of Cedar Lake Bog to energetic terms. Although he had barely sketched the outlines of this new energetic approach to ecology, its explanatory power seemed obvious enough. Why are trophic levels in ecosystems usually limited to four or five? In 1927 Charles Elton had proposed an answer to this intriguing question in terms of the relative size and numbers of various animals. For Lindeman a more elegant and fundamental explanation could be provided in terms of energy. Critics later complained that Lindeman's followers were "obsessed with calories."[82] But during the postwar decades the study of energy flow appeared to be of fundamental importance for understanding ecosystems.

Theoretical Foundations for a New Ecology

Arthur Tansley proposed the ecosystem concept as a more acceptable substitute for Frederic Clements's idea of the community as a complex organism. But neither Tansley nor Lindeman completely severed the ties between the older and newer concepts. With some qualifications, the idea of succession as a developmental process was an integral part of Lindeman's argument. Even more important, he borrowed Hutchinson's idea that energy and material transfer within aquatic systems constituted a kind of "metabolism." These were useful analogies, but they were no more than that. Like Hutchinson, Lindeman used organismal analogies as heuristic devices in developing theories. The mathematical theory, not the similarities between organisms and lakes, was truly important. The ecosystem concept was well-suited for this new type of study. Once the ecologist had delimited the system, the movements of energy and material within it could be measured.

Lindeman criticized earlier ecologists for taking a too biological perspective on their work. In important ways Lindeman removed the traditional biological orientation from ecology. This shift is particularly striking when one compares the new approach to trophic dynamics with the older ideas of Elton. Elton's ideas of food chains and niches were firmly rooted in natural history; he always seemed to have specific animals in mind when he discussed these theoretical units. But in the newer approach, species and populations seemed to evaporate. When considering biogeochemistry, V. I. Vernadsky had claimed, "the single living organism recedes from view; the sum of all organisms, *i.e.* living matter, is what is important."[83] Hutchinson and Lindeman took much the same view as the influential Soviet geochemist. Trophic levels did not correspond exactly with populations or species; a given species might simultaneously act as both herbivore and carnivore. From the perspective of the ecosystem this made little difference. In theory, a species's metabolic activities could be apportioned among several different trophic levels. Following this approach, however, trophic levels ceased to be truly biological units; instead, they became black boxes, idealized mechanisms for channeling and dissipating energy.

From one perspective Lindeman had deemphasized biology, but from another perspective ecosystem ecology became more biological than its predecessors. The split between botanists and zoologists was a prominent feature of pre–World War II ecology. Bacteria, fungi, and other "lower forms" of life were barely considered. Frederic Clements and Victor Shelford had made an initial attempt at synthesis in their 1939 book, *Bio-Ecology*. But ecologists of a younger generation found the book uninspiring. Lindeman's emphasis on energy flow focused attention on the unity of the ecosystem. Ecology was about more than animals or plants in isolation; it was about the interactions of various groups. Producers, consumers, and decomposers all played important functional roles within the ecosystem. For ecologists after World War II, the ecosystem promised a means for unifying what had previously been a discipline sharply divided along the boundary between zoology and botany.[84]

How important can a single paper be in establishing a new scientific specialty? Unlike other ecologists who studied productivity and trophic relationships, Lindeman placed his ideas within a coherent conceptual framework. Discussing his research within the context of the ecosystem allowed Lindeman to generalize rather broadly. The paper was not simply about lakes; it applied equally to all ecosystems, terrestrial or aquatic. Many of Lindeman's ideas were speculative, but this in itself encouraged constructive criticism. Finally, the drama surrounding the publication of the paper and Lindeman's tragic death focused considerable attention on his work. For all these reasons, "The Trophic-Dynamic Aspect in Ecology" became a catalyst for the development of ecosystem ecology. As we see in chapter 6, a number of social factors also encouraged this development.

6

Ecology and the Atomic Age

Actually, the atomic age can well provide the means of solving the very problems it creates.

—EUGENE P. ODUM, "Ecology and the Atomic Age"

Atomic energy today is still largely military; unfortunately, it has proved to make much better weapons than plowshares.

—EUGENE P. ODUM, "Radiation Ecology at Oak Ridge"

"THE AGE OF ECOLOGY began on the desert outside Alamogordo, New Mexico, on July 16, 1945, with a dazzling fireball of light and a swelling mushroom cloud of radioactive gases."[1] There is an element of truth in this provocative claim made by Donald Worster in the epilogue of his fine book, *Nature's Economy*. The potential dangers to health and environment posed by atomic energy were quickly recognized and eventually served as a primary target of popular environmental movements. In response, professional ecologists effectively used concerns over atomic energy as a convincing justification for ecosystem studies. But the "Age of Ecology" and the "Atomic Age" coincide for important reasons other than that suggested by Worster. Atomic energy also provided ecologists with an exciting new set of tools, techniques, and research opportunities. Thus, for the professional ecologist of the postwar period, nuclear power appeared to be a kind of double-edged sword: capable of wreaking environmental havoc, but also capable of unlocking many of nature's secrets for human benefit.[2] Perhaps even more important than these technical innovations were the economic opportunities of the atomic age. In the decades after World War II, the Atomic Energy Commission (AEC) became an important source of funding for ecological research. Ecosystem ecology clearly benefited from its association with one of postwar Washington's high-profile agencies, but the relationship was by no means asymmetrical. As Eugene Odum suggested in 1965, during the decades following World War II a feedback loop developed between nuclear power and ecology.[3] One might also think of a form of symbiosis developing between atomic energy and ecosystem ecology—a relationship in which both partners benefited. This chapter traces the development of that relationship.

The Metabolism of Ecosystems

As the United States entered World War II, Raymond Lindeman presented ecologists with a promissory note. His measurements of productivity and trophic efficiency were only rough estimates, but he had shown that such measurements could be made. By doing so the ecologist could discover something important about how ecosystems function. This promissory note was cashed a decade later by two brothers on a tiny speck of land a quarter of the way around the globe from New Haven. Eugene Odum and Howard Odum (figure 9) were not the only ecologists interested in studying energy flow in ecosystems after World War II; however, their study of a coral reef on Eniwetok Atoll appeared first, and it won the prestigious Mercer Award from the Ecological Society of America. Thus, it became an important exemplar for the new type of functional study of ecosystems that became so popular during the 1950s.

Eniwetok Atoll is a collection of some thirty tiny islands protruding from a horseshoe-shaped coral reef surrounding a shallow lagoon.[4] A distance of roughly twenty-five miles separates a wide, shallow inlet at the southern end of the horseshoe from the northernmost island at the top. During World War II, Eniwetok was an important military base, first for the Japanese and later for the United States. After the war, it became a site for testing nuclear weapons.[5] The largest island in the atoll, Eniwetok, served as a base for air support during the earliest postwar tests carried out during the summer of 1946 on Bikini Atoll, two hundred miles to the east. Later Eniwetok Atoll itself became a major testing site. Operation Sandstone, a series of tests begun in 1948 on a new generation of more efficient fission bombs, was carried out on islands in the northeast quadrant of the atoll. And in 1952, as part of Operation Greenhouse, the first thermonuclear device was detonated on the atoll. The 10.4 megaton blast obliterated the island of Elugelab, vaporized millions of gallons of ocean water in a fireball three miles in diameter, and gouged a crater half a mile deep and two miles long in the reef.[6] Atmospheric testing of nuclear weapons in the Marshall Islands continued throughout the decade and culminated in a series of thirty-three explosions during the summer of 1958.[7]

With the advent of nuclear testing, the Marshall Islands became a center for extensive biological, geological, and oceanographic research. Concerned with the potential effects of radiation on biological systems, a task force of scientists drawn from universities, private research institutes, and government agencies completed an initial survey of Bikini and neighboring atolls in 1946.[8] Subsequent studies were carried out during the period of atmospheric testing. As part of this research effort, the Atomic Energy Commission approached Eugene Odum to undertake a detailed ecological study of Eniwetok.[9] Odum, a professor of zoology at the University of Georgia, was already under contract with the AEC to do ecological research at the Savannah River nuclear reservation. Together with his younger brother Howard, an ecologist at the University of Florida, Odum spent six weeks in 1954 studying a reef on the windward side of the atoll. By time they arrived on Eniwetok, the atoll was hardly a pristine natural environment. Indeed, it was sufficiently radioactive that the Odums produced an autoradiographic image of a coral head simply by placing it on photographic paper. In a bit of understatement, the two ecologists introduced their report with the statement that the ongoing nuclear testing at Eniwetok provided a unique opportunity "for critical assays of the effects of radiations due to fission products *on whole populations and entire ecological systems in the field.*"[10]

Although neophytes in radiation biology and almost totally unfamiliar with the ecology of tropical reefs, the Odum brothers were better prepared than most ecologists for the type of project proposed by the AEC. Both were already working on major ecosystem studies in the United States. Beginning in 1951 Eugene had initiated what became a long-term study of succession and productivity on farmland abandoned during the construction of the Savannah River nuclear facility. Howard was in the midst of an even more ambitious study of warm mineral springs in Florida.[11] Although his four-year study of Silver Springs was published two years after the Eniwetok study, by 1954 he had already perfected a number of techniques that the brothers used to study the metabolism of the coral reef. A student of G. Evelyn Hutchinson, Howard Odum shared his teacher's penchant for making bold theoretical speculations and then testing their consequences with equally bold experimental studies. Eniwetok provided Howard with a perfect opportunity for exploring not only Lindeman's theoretical claims but also some of his own controversial ideas about the nature of ecosystems. It is also significant that the Eniwetok study came a year after the publication of Eugene's *Fundamentals of Ecology* (1953). The chapter on energy, written by Howard, outlined general principles and provided a few examples. These examples, however, were constructed from miscellaneous data collected from several unrelated sources. Prior to 1954, no one had investigated a complete ecosystem with the intent to measure its overall metabolism. Eniwetok and the Atomic Energy Commission provided the opportunity to do so.

The Odums have often claimed that ecosystem ecology is predicated upon an antireductionistic, "whole-before-the-parts" philosophy,[12] but the Eniwetok study actually demonstrated how readily the Odums moved among levels of organization. In reality, the parts and the whole were equally important, and both were investigated simultaneously. The study was holistic in the sense that its primary focus was the overall structure and function of the ecosystem. The metabolism of the entire reef was most important, not the metabolism of constituent individuals or populations. In fact, the Odums were so unfamiliar with corals that they could not identify most species on the reef.[13] In a very real sense, the reef at Eniwetok Atoll was a "black box" whose total inputs and outputs of energy were being measured.[14] But understanding the whole could not be accomplished without understanding at least some of the parts. The Odums may not have known the scientific names of the corals, but they were interested in the functional role played by this important component of the ecosystem. Indeed, the peculiar nature of coral held the key to understanding the system as a whole.

In antiquity corals were considered plants, but zoologists, recognizing them to be coelentorates, had long claimed them for their own. Early in the twentieth century it was discovered that coral polyps often contained symbiotic unicellular algae. Functionally, if not taxonomically, the polyp had to be considered "part plant."[15] For the Odums this was not merely a curiosity; it suggested a pervasive symbiosis. Existing in the inert calcium carbonate skeleton surrounding the coral polyps were networks of filamentous green and blue-green algae. These were sufficiently dense that the entire coral head often had a distinctly greenish hue. Previous investigators had dismissed the filamentous algae as parasites that actually weakened the coraline skeleton. But the Odums suggested an alternative explanation: mutualism. Although they could not actually demonstrate a transfer of nutrients, the ecologists proposed that the algae shared the products of photosynthesis with the coral polyps. In exchange, the polyps protected the delicate filamentous algae from browsing predators, intense sunlight, and other environmental hazards. Some indirect evidence supported this claim. Autoradiographs demonstrated that the radioactivity from the water was restricted almost entirely to the polyp zone, barely penetrating to the algae beneath the polyp. Where living polyps were absent in the coral head, however, the embedded algae were intensely radioactive.

The coral head could be considered a kind of ecosystem within an ecosystem. The amount of plankton in the water flowing over the reef provided insufficient food to support the coral animals. Thus, the metabolic economy of the head depended upon the photosynthetic production of symbiotic algae. The complex structure of the head provided a mechanism for recycling basic nutrients to these producers; the small leakage of nutrients escaping from the system was just balanced by nutrients entering the reef from the ocean. An earlier survey of Eniwetok Atoll had suggested that the entire reef might also be self-supporting. Compared to the reef, the open ocean was not very productive; it probably was not a major supplier of organic material for the coral. This finding was, however, inconclusive. Put simply, there were two possibilities. Perhaps the reef was a highly efficient filtering mechanism, and even though there was little organic material in the ocean water, the large volume of water passing over the reef might yield a significant food supply for coral polyps. In economic terms, this scenario suggested that the ecosystem was operating at a deficit, sustained by energy and nutrients imported via the oceanic current. Alternatively, the community might be truly self-sufficient, recycling necessary nutrients within itself and living off the photosynthetic production of its algae. Choosing between the alternative explanations required the construction of an energy budget for the reef, as a whole. If productivity did not at least equal respiration, then the community was not self-sustaining.

Water samples from the front and back of the reef contained almost the same amount of nutrients. This suggested that the coral was not simply filtering food imported from another ecosystem. More important, using an ingenious technique perfected by the younger

Odum, the two ecologists demonstrated that the productivity and respiration of the reef were nearly equal. Howard Odum's "diurnal flow" method rested upon the notion that the reef was like a river; water flowed from "upstream," at the front of the reef, to "downstream," at the back of the reef.[16] Simultaneous determinations of dissolved oxygen in front of the reef and behind could be used to measure the overall metabolism. An increase in oxygen indicated that rate of photosynthesis was greater than the rate of respiration. A decrease indicated that respiration was occurring at a more rapid rate than photosynthesis. Periodic measurements of dissolved gases taken during a twenty-four hour period could be combined to form an energy balance sheet for the living community. Oxygen determinations taken at night, when no photosynthesis occurred, provided a measurement of the rate of community respiration. Those taken during the day, when both processes occurred, provided a measurement of net primary production. Adding community respiration (determined at night) and net primary production (determined during the day) gave the total or gross primary production for the ecosystem. Using these determinations, the complete energy budget for the reef could be estimated. Expressed in terms of grams of glucose/meter^2/day, the Odums's calculations indicated that the energy income of the reef was slightly higher than its losses. Although the energy budget was slightly out of balance, indicating that the community actually produced slightly more than it consumed, the Odums admitted that the methods used were not sufficiently refined to support this conclusion with any degree of confidence. Indeed, they favored the view that the reef was a true steady state ecosystem or in more traditional ecological terms a climax community.[17] This did not mean that the reef was a static system—far from it. The visible biomass, or standing crop, represented only a small fraction of the energy captured by producers. The Odums estimated that there was a complete turnover of biomass more than once a month on the reef. As the Harvard ecologist George Clarke had earlier suggested, the ecosystem might be likened to a factory filled with spinning wheels.[18] But it was a rather strange factory, for virtually all the ecosystem's production was used to maintain the machinery. Little was stored as visible biomass.

The Eniwetok study was a landmark in ecological research, important to both individual researchers and the discipline of ecology as a whole. The reef with its close symbiotic relationships between coral and algae was an excellent example of a highly structured, self-regulating system—a nascent view of ecosystems toward which both

Odums were strongly attracted. Over millions of years, the Odums reasoned, the reef had evolved an optimum composition of interacting species, a composition that favored the stability of the whole. According to this "stability principle," natural selection amounted to "survival of the stable."[19] By favoring stability, natural selection perfected self-regulating interactions in ecosystems. This belief in ecosystem stability, a belief shared with G. Evelyn Hutchinson and several other ecologists, is discussed in greater detail in chapter 7. Suffice it to say that the experience at Eniwetok focused and reinforced ideas that the Odums already held about the fundamental nature of ecosystems. For Eugene, the coral reef became an exemplar for self-regulating, self-maintaining systems.[20]

If Eniwetok was an important milestone in the careers of Eugene Odum and Howard Odum, then it was equally significant as an indicator of where post–World War II ecology was heading. The Mercer Award has often gone to ecologists whose research establishes new trends. The Odums's paper, which won the award in 1956, was the first of several similar studies. It was followed two years later by Howard's monumental study of Silver Springs, certainly the most ambitious ecosystem study of the 1950s. Supported by a $20,000 grant from the Office of Naval Research (ONR)—a substantial sum for that time—the younger Odum had spent four years studying the flow of energy and materials in this aquatic system. During the course of this work, Howard not only developed the techniques used at Eniwetok, but he also produced the most detailed account of the energy flow and nutrient cycling in an ecosystem. It remains the ecosystem study most commonly referred to in undergraduate textbooks. Other studies during the 1950s by ecologists such as John Teal, Frank Golley, Lawrence Slobodkin, and Edward Kuenzler also helped to establish energy flow as a central area of research within the discipline.

The case of John Teal is particularly significant for what it reveals about the emergence of a self-conscious group of ecosystem ecologists. Teal, a graduate student at Harvard University, had begun physiological research using leeches. As he later recalled, he quickly became bored with his rather traditional laboratory project. After reading Lindeman's trophic-dynamic paper, probably in 1952, Teal became excited about studying the "metabolism" of a more complex biological system: a small cold spring near Concord, Massachusetts.[21] Like the Odums, who were beginning their research projects at the same time, Teal wanted to make a more precise functional analysis of an ecosystem than Lindeman had been able to do. Such detailed studies of energy flow, Teal hoped, would allow ecologists to understand ecosystems in the same way that traditional metabolic studies allowed laboratory physiologists to understand how individual organisms functioned.[22] The system that Teal chose to study was ideal for the purpose. Only two meters in diameter, the spring was as close as nature could come to being a "laboratory ecosystem."[23]

During his two-year study, Teal was unaware that several other ecologists were doing much the same type of research. Although he had been introduced to Lindeman's paper by his adviser, George Clarke, he did not discuss it with either Clarke or other ecologists. Only after completing the field work did he meet Howard Odum, whose influence can be seen in the diagrammatic presentation of Teal's data. By 1957, when Teal's study and Odum's Silver Springs paper appeared in *Ecological Monographs,* an informal network of ecosystem researchers was beginning to form. On Howard's recommendation, Teal was hired as a postdoctoral fellow by Eugene Odum, who was building a marine laboratory on the Georgia coast. Together with several other young Ph.D.s and graduate students Teal established the laboratory on Sapelo Island and began one of the first ecosystem studies of fragile and environmentally important salt marshes.

New Economic Opportunities

The Eniwetok study served as an early benchmark for ecosystem studies, but for the historian it also serves as a symbol for post–World War II science. Academic science in America was revolutionized by the war. Scientists had been mobilized on a large scale, and they had contributed decisively to winning the war. For scientific planners facing the uncertainties of the post–war world, military strength, economic growth, and human welfare seemed to depend upon this new partnership between the state and the scientific community.[24] For individual scientists, including ecologists, the postwar period was also a watershed. If, as Robert McIntosh suggests, the classical ecologist was a rugged individualist working with a few pieces of string and a pH meter,[25] the image of the new ecologist was quite different. Not only did "fancy instruments with flashing lights and clicking sounds" become part of the ecologist's tool kit,[26] but in the new environment of government contracts and grants, the successful scientist had to combine the skills of investigator, entrepreneur, and bureaucrat.

For ecosystem ecologists this new environment offered many opportunities for building research programs. During the period from

1950 to 1965, the "golden age" of funding, several government agencies vied for the opportunity to support innovative basic research in biology.[27] Even in agencies such as the Office of Naval Research, which funded Howard Odum's Silver Springs study, there was strong commitment to supporting innovative research regardless of its direct military application. Recalling his role in setting research priorities at ONR, the oceanographer Roger Revelle noted:

> Helping the Navy was not a good reason for doing research. The only good reason for doing research was that they [scientists] wanted to do it in their bellies; they were driven by curiosity, the desire for discovery and the desire for fame, which is what drives scientists. . . . In fact, we had two or three different mottos in my part of the ONR. One was that any proposal for less than $5,000 we automatically funded. Another was, as I say, that anybody who said he wanted to do this because it was good for the navy we automatically turned him down, unless it was for less than $5,000.[28]

Revelle's recollections, forty years after the fact, may be slightly romanticized. But throughout the 1950s there was a sense of excitement in many funding agencies; there seemed to be unlimited opportunities for supporting basic science, and great things seemed possible as a result of this new partnership between science and government.

No federal agency epitomized this attitude as well as the Atomic Energy Commission. Beginning in 1948, two years after the agency was established, its Division of Biology and Medicine began supporting a diverse program of research in the life sciences.[29] Support for ecology never rivaled that for genetics or physiology, but impressive research programs were started at a few universities and national laboratories located at Oak Ridge, Brookhaven, Savannah River, and Hanford. Much of this research was directed toward applied problems of radiation ecology, an area of research that had gained limited support even during the war.[30] However, as the case of Eugene Odum at the University of Georgia illustrates, ecologists interested in basic science could also benefit from the largesse of the Atomic Energy Commission. Combining the opportunities provided by the Savannah River nuclear facility with other unique local sources of funding, Odum created one of the most influential programs in ecology. This was done within the structure of a small state university, an institution that otherwise would have provided a poor environment for building such a program.

The Savannah River nuclear facility was built as a result of President Truman's decision in 1950 to begin production of the hydrogen bomb.[31] The primary function of the plant was to produce tritium for the new bomb, but plutonium was also to be produced there. Construction began on the site in Aiken, South Carolina, in the spring of 1951, and the first of five reactors began operating two years later. Because the Savannah River facility was an entirely new facility, it provided a unique opportunity for doing pre-installation environmental surveys; that opportunity had not been considered during the wartime push to build the first atomic weapons.[32] Early in 1951, the AEC invited proposals for ecological research from the Universities of Georgia and South Carolina.

Eugene Odum submitted a detailed proposal for an ambitious environmental survey of the terrestrial and aquatic ecosystems surrounding the Savannah River facility. The initial survey would be followed by subsequent studies once the nuclear reactors were operational. Thus, Savannah River would serve as a kind of large-scale experiment to determine the effects of emitting low levels of radiation and large amounts of heated water into the environment. Odum's proposal called for six full-time principal investigators supported by graduate research assistants. The initial annual budget for the project, to be shared by the AEC and the university, was estimated at slightly more than $267,000.[33]

Two decades later a grant proposal of this magnitude would not have raised eyebrows, but in 1951 the AEC was not contemplating an environmental impact statement. Odum's proposal was rejected, and in a subsequent meeting with AEC officials he was informed that the commission was prepared to provide a yearly grant of approximately $10,000 to prepare an inventory of the animal populations in the area. A similar amount would be given to the University of South Carolina for botanical work. Odum was able to convince the commission to use some of the money to support a study of secondary succession on the abandoned farm land surrounding the plant. He also expressed interest in experimenting with radioactive tracers, a technique he knew little about but one that he and his students later used extensively.

As Odum recalled, enthusiasm for the project quickly declined on campus when it was learned that the university would not receive a $150,000 "sugar plum" from the government.[34] Apparently this was also the case at the University of South Carolina, which dropped out of the project after only a few years. Odum, however, pushed ahead and submitted a much more modest proposal to the AEC. He also made a strategic decision that, in retrospect, insured the success of his research program. The limited money provided by the AEC would not be used to support senior researchers; faculty members would continue to be paid by the university. Instead, virtually the entire research budget was earmarked for graduate assistantships. As a result, Odum was able to attract a steady stream of students to his fledgling ecology program.

What began as a small and uncertain source of support, rapidly developed into a long-term relationship between the University of Georgia and the AEC. By 1960 the annual budget for ecological research at Savannah River had increased to $60,000, and the AEC had agreed to establish a permanent ecological laboratory at the site. Frank Golley, a young faculty member at the university, became the resident director of the laboratory. Golley, a recent graduate of Michigan State University, had studied energy flow through a single food chain in an old field community. His dissertation was perhaps the first attempt to study the metabolism of a terrestrial community. Odum, who had met Golley on a visit to Michigan State, was instrumental in bringing the young ecologist to the University of Georgia. The two shared not only an interest in the developing specialty of radiation ecology but also a broader commitment to ecosystem ecology. Golley was, therefore, the perfect choice to direct the expansion of the ecological research program at Savannah River.

One can only speculate about how successful Eugene Odum's research program would have been without the long-term support of the AEC. It seems safe to say that the University of Georgia would not have become such a major center of ecological research were it not for atomic energy. But the Atomic Energy Commission was only part of the equation for success. In contrast to the Oak Ridge and Hanford laboratories, ecological research at the University of Georgia never became narrowly identified with radiation ecology. To a large extent, this was owing to the fact that Odum, Golley, and other ecologists at the university had a much deeper commitment to ecosystem studies than to radiation ecology per se. Contract work at Savannah River was viewed, not as an end in itself, but as a means to help subsidize the broader ecological program at the university.[35] The subsidy was more than just economic. The laboratory at Savannah River provided research opportunities for graduate students and served as a kind of training base for young Ph.D.s. Those who made good were often brought back as faculty members on campus. Geography helped too; being separated from the university campus by more than one hundred miles, the Savannah River operation could be maintained as a semi-autonomous part of the larger research program. Finally, although funding from the Atomic Energy Commission was crucial, it was not the only source of support for the ecology program that Odum was struggling to build during the 1950s.

Figure 1.
Henry Chandler Cowles, 1913 (negative no. DN 60,959, Chicago Historical Society).

Figure 2.
Frederic Clements, 1914 (American Heritage Center, University of Wyoming).

Figure 3.
Charles Elton (*center*) with Aldo Leopold (*right*) and the Canadian ornithologist William Rowan (*left*) at the Matamek Conference on Population Cycles, Labrador, 1931 (University of Wisconsin Archives).

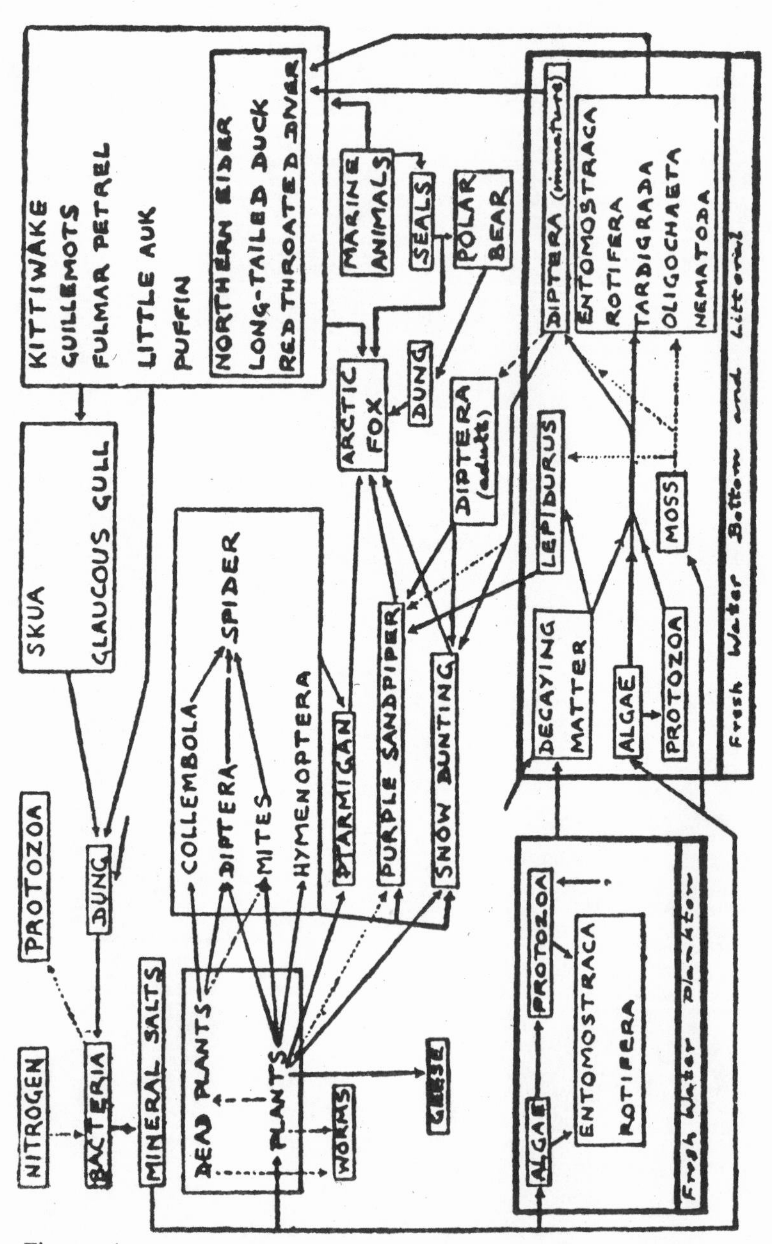

Figure 4.
Food cycle of the animal community on Bear Island, near Spitsbergen (Charles Elton, *Animal Ecology*, Sidgwick and Jackson, 1927). Elton sometimes referred to this diagram as a "nitrogen cycle."

Figure 5.
G. Evelyn Hutchinson at the Osborn Memorial Laboratory 1939 (Yale University Archives, Manuscripts and Archives, Yale University Library).

Figure 6.
Sketch of Sir Arthur Tansley by Jane de Glehn, 1939 (by permission of the President and Fellows of Magdalen College, Oxford).

Figure 7.
Raymond Lindeman,
ca. 1939
(University of Minnesota Archives).

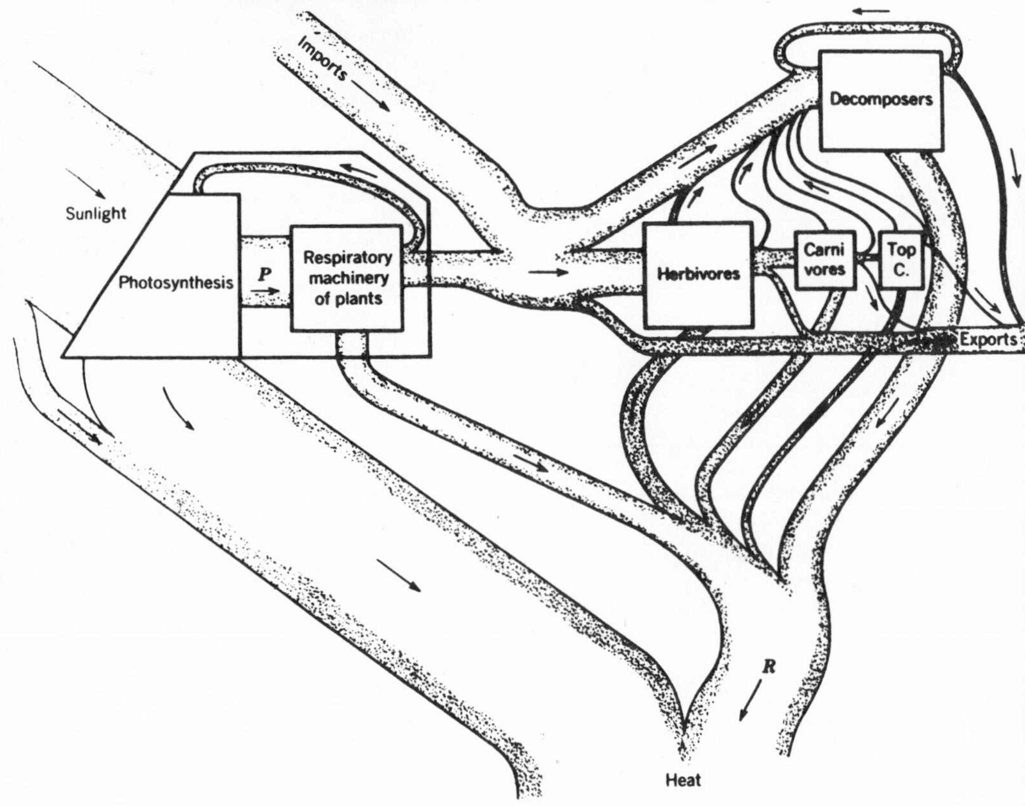

Figure 8.
Simplified diagram of energy flow through Silver Springs, Florida (Howard T. Odum, *Environment, Power, and Society*, John Wiley, 1971).

Figure 9.
Eugene and Howard Odum receiving the 1987 Crafoord Prize. *Left to right*: Mrs. Elizabeth (H. T.) Odum, Mrs. Martha (E. P.) Odum, Howard T. Odum, Mrs. Anna-Greta Crafoord, Eugene P. Odum, King Carl XVI Gustaf (Royal Swedish Academy of Sciences).

Figure 10.
Gene Likens (*left*) and F. Herbert Bormann (*right*) at Hubbard Brook (F. Herbert Bormann).

Eugene Odum had a knack for taking advantage of regional opportunities. Trained as an ornithologist, Odum and his students used Sapelo Island, a marshy piece of land on the Georgia coast, as a study area. The island was owned by R. J. Reynolds, Jr., who used it as a hunting preserve. During the course of his visits to the island Odum got to know the tobacco heir, and in 1954 Reynolds donated Sapelo to the state of Georgia and established the Sapelo Foundation to support ecological research on the island. Like Savannah River, research on Sapelo Island began modestly. John Teal, one of the first ecologists to arrive, found that the "laboratory" was an old barn with a scale, microscope, and few other pieces of equipment.[36] The laboratory did provide access to a nearly pristine salt marsh, however, and the island continued to draw graduate students and young Ph.D.s eager to do ecosystem research in this unique natural environment. The Sapelo Foundation might not have provided the level of funding that the AEC did, but Odum used the money in the same way to expand graduate and postdoctoral research at the University of Georgia.

Looking back on the origins of his program, Odum noted the irony of a major center of ecological research based at a small university in Georgia.[37] To a certain extent, his success as a scientific entrepreneur was fortuitous. Savannah River and Sapelo Island provided Odum with unusual opportunities unavailable to most other ecologists. But Odum skillfully built upon this foundation. By the early 1970s his Institute of Ecology was attracting several million dollars in support from a wide variety of federal, state, and private agencies.

During the early 1950s Odum was unique among ecologists in the level of support that he received from the AEC. This situation began to change around the middle of the decade. In 1955 John N. Wolfe, an ecologist at Ohio State University, became an administrator in the Division of Biology and Medicine.[38] Three years later he was promoted to chief of the newly established Environmental Sciences Division at the AEC. Under Wolfe's energetic leadership the scope of AEC, funding in ecology greatly expanded, particularly support for university-based research. Just as significantly, during this period ecological programs at the national laboratories expanded, particularly at Oak Ridge.

As Chunglin Kwa has shown, despite its similarities to the University of Georgia program, the development of the Oak Ridge ecology program was constrained by its unique setting.[39] In 1954, Stanley Auerbach was brought to Oak Ridge to develop research on the movement of radionuclides in the environment. Auerbach's position was more tenuous than Eugene Odum's. The Oak Ridge National Laboratory (ORNL) was strongly oriented toward the physical sciences, and it was headed by the egocentric Alvin Weinberg. Weinberg had little interest in biology, although he believed that environmental studies might play a small, but significant, part in his scheme for making the laboratory a center for "big science." Despite these institutional constraints, Auerbach skillfully built an impressive ecological program. Ecologists at Oak Ridge often held joint appointments in biology departments at the University of Tennessee. This provided access to students and gave the program an academic dimension. Auerbach also used his position as secretary of Ecological Society of America to raise the visibility of the program in professional circles. Intellectually Auerbach drew heavily upon the ideas of Eugene Odum, who served as a consultant to the Oak Ridge program, but in important ways the two programs were different, a reflection of the influence of the physical sciences at the Tennessee laboratory. By the late 1960s Oak Ridge was a leading center for the study of radiation ecology, and its ecologists were pioneering the use of computer simulation, which eventually became the subspecialty of systems ecology. During the early 1970s the Oak Ridge group played a central role in organizing the large-scale ecosystem studies associated with the International Biological Program (IBP). These later developments are discussed more fully in chapter 9.

New Tools for the Ecologist

The Atomic Energy Commission provided crucial financial resources for the institutional development of ecosystem ecology. But the atomic age also provided ecologists with an array of exotic new techniques for their research.[40] Some had relatively limited applications. Eugene Odum and others experimented with the use of radioactive "tags" for tracking the movements of animals and estimating population densities. As director of the new Geochronometric Laboratory at Yale University during the early 1950s, Edward Deevey began using radiocarbon dating in his analysis of pollen in sediments of Linsley Pond. This provided Deevey with a more precise technique for attacking the paleoecological problems that he had begun to study in the late 1930s.

Radiation could also be used to study the effects of stress on living systems. Given public concerns about fallout, radioactive waste storage, and nuclear war, it is not surprising that radiation effects became one important focus of ecological research during the 1950s and 1960s. Not only were the radiosensitivities of individual organisms studied but also those of whole ecosystems. In a number of cases, natural areas were irradiated to determine damage and recovery of constituent species after exposure to radiation.[41] For example, two sites at the Brookhaven National Laboratory were exposed to high levels of radiation for a period of six months. In both cases, a forested area and an old field, a source of gamma rays (^{137}Ce or ^{60}Co) was stored in an underground lead-shielded container. The radiation source could then be raised or lowered by remote control. The Brookhaven study and similar experiments in Georgia and Puerto Rico demonstrated that at least some species of plants were relatively sensitive to radiation damage. It also suggested that succession, food chain dynamics, and diversity might be adversely affected by chronic exposure to even fairly low levels of radiation.[42]

For most ecologists involved with atomic energy, however, the most exciting application of radiation was in tracer studies. For ecosystem ecologists radioactive tracers appeared to be a powerful tool, one that held great promise for unraveling the complex internal processes of the ecosystem. "Radioactive tracers have already been well exploited by the physiologist," noted Eugene Odum in 1959, "but the ecologist is just beginning to develop techniques for studies in 'community metabolism.'"[43]

Radioactive tracers revolutionized many areas of biological research after World War II.[44] Even before the war, radioactive isotopes had been widely used as tracers to study the metabolism of plants and animals. Because radioactive isotopes are readily detectable and have the same chemical properties as their nonradioactive analogues, the use of tracers transformed the study of metabolic pathways. During the 1930s this new physiological methodology was widely discussed in general scientific periodicals such as *Nature* and *Science,* and it is perhaps not surprising that ecologists such as G. Evelyn Hutchinson, who were interested in the "metabolism" of ecosystems, soon began experimenting with radiotracers. But prior to World War II, the availability of isotopes was limited. In 1941, Hutchinson's initial attempt to study the phosphorus cycle in Linsley Pond using a radioactive tracer (^{32}P) failed when the cyclotron at Yale produced only half the amount of isotope required for the experiment.[45] After World War II the situation changed dramatically. Nuclear reactors provided an almost unlimited supply of radioactive isotopes for biological research. The distribution of these isotopes for research became the centerpiece of the government's Atoms for Peace program, and AEC officials actively promoted the use of this new technology.[46] For ecologists, the buffer zones around nuclear facilities provided large natural areas where radioactive tracers could be used for ecosystem experimentation. University researchers such as Hutchinson and his students successfully exploited the use of tracers. But given the close proximity to sources of radionuclides and the availability of large, isolated areas for field research, it is not surprising that the ecologists at the nuclear reservations at Oak Ridge, Hanford, Brookhaven, and Savannah River were at the forefront of the postwar development of ecosystem tracer studies.

Radioactive tracers provided ecologists with a means for quantifying the movement of materials and energy through the ecosystem. This new technique was particularly useful for studying biogeochemical cycles in aquatic ecosystems. The radioactive isotope of an essential element such as phosphorus could be added directly to the water. The movement of the isotope through the various trophic levels of the community, and between the community and the abiotic environment, could then be monitored with a radiation detector.[47] This added a new dimension to biogeochemical studies, for it provided the ecologist with a much clearer picture of the dynamics of the cycle. By following the movement of the tracer from one compartment of the ecosystem to another, the ecologist could more accurately estimate the rates at which these movements occurred.

While tracers could be used to measure directly the movement of materials through the ecosystem, they could also be used to measure indirectly the flow of energy. Plants, the producers in the ecosystem, were labeled with a radioactive isotope. At subsequent time intervals the various consumers in the community were then sampled for radiation. This type of study yielded two important types of information about the internal workings of the ecosystem. First, it could be used to isolate individual food chains. If a particular species of plant was labeled, then only herbivores feeding on the plant and the carnivores feeding upon those herbivores would later become radioactive. Second, tracer studies could be used to answer the question: How long does it take for energy to move through the ecosystem? By continuously monitoring the various animal populations for radiation, one could estimate the time required for elements originally in the plants to reach the end of a food chain. In short, ecologists used tracers in much the same way that biochemists and physiologists did, but on a much larger scale. By following the movement of tracer elements, the ecologist could accurately determine the actual pathways of energy flow, the rates at which the energy flowed, the amount of time that various elements remained within particular compartments of the ecosystem (residence time), and rates at which these elements moved from one compartment to another (turnover rate).

Tracer studies worked best in small, isolated ecosystems populated by relatively sedentary organisms. G. Evelyn Hutchinson's favorite study site, Linsley Pond, was a perfect place for such research. After the war, when a reliable supply of radioactive phosphorus could be procured from the Oak Ridge nuclear facility, he and his graduate student Vaughan Bowen were among the first to publish isotopic studies of phosphorus cycling. Even more ideal systems could be created in the laboratory. Some of the neatest data on the movement of phosphorus came from Robert Whittaker's aquarium studies at the Hanford laboratory.[48] In these microcosms, the radioactive isotope could not escape, and the fate of all the phosphorus added to the system could be accurately determined. But other ecologists were not deterred by the technical problems posed by complex natural ecosystems. The popularity of tracer studies is evident by the large number of papers on biogeochemical cycles and energy flow presented at national symposia on radiation ecology.[49]

Justifying Ecosystem Research

If atomic energy provided new research opportunities for ecosystem ecologists, then it also provided a convincing justification for their new specialty. Despite government efforts to promote "Atoms for Peace" during the 1950s, most Americans continued to associate atomic energy with bombs.[50] Latent fears about the effects of fallout and waste disposal fueled an undercurrent of uncertainty that occasionally flared into public controversy. Although ecologists could offer no panaceas, they could hold out the promise that further research might provide solutions to the environmental problems posed by atomic energy. At the forefront of this movement was Eugene Odum who did much to publicize the environmental effects of radiation in his popular textbook, *Fundamentals of Ecology*.

Fundamentals of Ecology changed in some significant ways as it went through three editions between 1953 and 1971. The most striking change in the second edition was the addition of an entire chapter on radiation ecology. By 1959 when the new edition appeared Odum had developed strong ties with the AEC, but there was nothing in his rather traditional academic training that had prepared him to be a radiation ecologist. His introduction to the subject came in 1957, when he was awarded a Senior Postdoctoral Fellowship from the National Science Foundation.[51] During his fellowship year Odum spent time at the Nevada Proving Grounds and at the nuclear facility at

Hanford, Washington. This experience provided him with the background to write the chapter on radiation ecology.

Odum's discussion of radiation ecology was perhaps the first review of the subject directed toward a general audience. Its appearance was quite timely. By the mid-1950s there was widespread public concern about the effects of low-level radiation in the environment. The United States, the Soviet Union, and Great Britain continued to test weapons in the atmosphere, and the controversy over "fallout" increased public awareness of radiation. Scientists were divided on the issue. The geneticist H. J. Muller, who was awarded the Nobel Prize in 1946 for his demonstration of the mutagenic properties of x-rays, warned against "race poisoning" from nuclear tests.[52] According to Muller even an extremely small increase in the mutation rate would eventually lead to a significant "genetic load" on the human genome. Muller's claim was highly controversial, but other prominent geneticists also voiced concerns about potential dangers of fallout for individuals and populations. Ecologists also responded to the problem, but from a slightly different perspective. After Hiroshima and Nagasaki, the American public could easily imagine the deleterious genetic and physiological effects of radiation. Less obvious, but no less ominous, were environmental effects. Ecologists, therefore, could present radiation in the environment as a pressing social and scientific issue, one that was poorly understood. At the same time they could use the problem as a justification for government support of what they considered an exciting line of research.

In his chapter on radiation ecology, Odum discussed the fallout problem briefly, but minimized the general risk to human health or the environment. For Odum, two other problems were more worrisome: biological magnification and nuclear waste storage. Researchers at the Hanford nuclear facility had recently discovered that radionuclides discharged into the Columbia River in trace amounts sometimes became greatly concentrated as they moved through the food chain. For example, the concentrations of radioactive phosphorus (^{32}P) were often hundreds or thousands of times greater in vertebrates at the ends of aquatic food chains than in the cooling water leaving the nuclear reactors. The potential danger from biological magnification seemed obvious. As Odum pointed out, "we could give 'nature' an apparently innocuous amount of radioactivity and have her give it back to us in a lethal package!"[53] The phenomenon of biological magnification later became more closely identified by the public with the problem of pesticides in the environment. Rachel Carson's *Silent Spring*, published three years after the second edition of Odum's textbook, made this a cause célèbre for the popular environmental movement.[54] But Odum and other radiation ecologists were less alarmist in their attitude toward the biological magnification of radionuclides. The actual danger of biological magnification depended upon a number of variables including the half-life of the nuclide, the ability of organisms to retain the element, and the concentration of the nonradioactive isotopes of the element in the environment. Furthermore, the phenomenon of biological magnification opened up new opportunities for ecosystem research. Because certain radionuclides accumulated in the various trophic levels of an ecosystem, tracer studies could determine the flow of energy and materials through the system. Thus, Odum concluded, "Man's opportunity to learn more about environmental processes through the use of radioactive tracers balances the possible troubles he may have with environmental contamination."[55]

Disposal of radioactive waste was a far more troubling problem for Odum. Like the emotional debate about fallout, concerns about waste treatment and disposal had become a controversial public issue by the time Odum revised his textbook.[56] The Atomic Energy Commission had a twofold policy on waste disposal. For low-level waste, the policy was "dilute and disperse." Liquid wastes underwent preliminary treatment to reduce radioactivity, and then they were released into waterways. Solid wastes were buried or dumped into the ocean. The AEC was convinced that this method of disposal was both safe and environmentally harmless. Because there was no satisfactory way to dispose of high-level nuclear waste, the AEC policy was "concentrate and contain." By 1957 highly radioactive liquid waste from the nation's reactors amounted to sixty-two million gallons, most of which was stored in underground tanks at Hanford, Washington. This temporary storage was to be superseded at a later date when suitable means of permanent disposal were developed. But by the time that Odum wrote his chapter on radiation ecology the tanks were already beginning to leak.

A number of scientists publicly voiced concerns about both aspects of the Atomic Energy Commission's disposal policy.[57] Odum's chapter reflected these misgivings. He admitted that the nine billion gallons of low-level radioactive waste entering the environment each year constituted a mere "drop in the bucket" compared to the vast volume of the oceans.[58] Nonetheless, Odum warned that the environmental impact of low-level waste might become critical late in the century if nuclear power became a major source of energy. The proliferation of reactors and the economic incentives to minimize the cost of waste treatment would inevitably lead to greater environmental contamination. Moreover, the possibility of biological magnification meant that even the limited discharge of low-level wastes into waterways was not necessarily harmless.

The problems that Odum stressed in his chapter on radiation ecology were not the evolutionary, population-level problems raised by geneticists such as H. J. Muller. Rather, they were ecosystem problems involving complex interactions of both abiotic and biotic components. Odum concluded that these problems were not yet critical in 1959, but that they might become critical in the near future. Therefore, solving these problems served as a compelling justification for expanding ecological research, specifically the type of ecosystem studies that Odum had been pioneering. Some ecological interactions could be studied in the laboratory. Ultimately, however, the fate of radionuclides in the environment could only be understood through experiments on natural ecosystems. "In the not too distant future," Odum concluded, "the radioecologist may well be one of those who must help decide when to contain and when to disperse the waste materials of the atomic age. If the ecologist does not know what to expect in the biological environment, who will?"[59]

A Symbiotic Relationship

At the beginning of this chapter I characterized the relationship between ecosystem ecology and atomic energy as symbiotic. From an evolutionary viewpoint, symbiosis implies more than mutual benefit. The boundaries between cooperation and competition, parasitism and mutualism are often poorly marked and easily crossed. Altruism may not reflect good will, but simply self-interest in disguise. Social symbiosis is no less complex and ambiguous than its biological model. Such is the case with ecosystem ecology in the atomic age.

The nascent ecological specialty that emerged from World War II clearly benefited from the rise of atomic energy. As we have seen, radionuclides provided ecologists with a new set of exotic tools and new research opportunities. In the Atomic Energy Commission ecologists found a rich source of financial support, and in public concerns over the effects of radiation they found a convincing justification for their new lines of research. By embracing atomic energy were ecosystem ecologists motivated by nothing more than self-interest? And why should the atomic energy establishment have taken interest in a still rather insignificant scientific specialty struggling to establish itself?

Certainly one cannot entirely rule out environmental concerns as part of the answer to the second question. The AEC had a dual mandate; the agency was established in 1946 both to promote and to regulate the development of atomic power.[60] As critics pointed out, regulation usually took a back seat to promotion, but in its regulatory capacity, the agency relied upon the technical input from scientists, including ecologists. Even at the height of World War II, there had been concerns about the environmental impact of atomic energy. During the construction and early operation of the nuclear facility at Hanford, high officials of the Manhattan Project had authorized limited environmental studies.[61] And extensive biological and geological surveys were part of Operation Crossroads, the first postwar atomic tests at Bikini Atoll. Nonetheless, I think that it is safe to say that concern for the environment was never the principal rationale for supporting ecological research. If it had been, the AEC would probably have supported a complete preoperational environmental study of the Savannah River site in 1951, rather than providing Eugene Odum with only modest funds to survey animal populations near the new reactors. But if concern for the environment was only a small part of the equation, then what was the government's primary rationale for supporting Odum and other ecosystem ecologists?

In her stimulating book, *A Fragile Power,* Chandra Mukerji argues that the state is less interested in specific technical information that scientists generate than in their broader technical expertise.[62] The scientific community, in Mukerji's view, constitutes an elite reserve labor force. The state supports scientific research, even research with little apparent practical application. In return, the state gets a pool of highly trained problem solvers. Developing such a pool of experts was a major priority of the government immediately after the war.[63] Funding basic research in ecology may have supplied a means for generating technical information and allowed ecologists to pursue research they found exciting, but in addition it provided the government with the means for training a cadre of experts knowledgeable about ecology and the environmental effects of radiation.

There is more to Mukerji's thesis than this, for she claims that the state uses science to legitimate its policies. One need not accept this thesis in all its details to apply it to the case of atomic energy. Ecologists and other scientists provided the AEC with technical information about the effects of atomic energy. But this information was often ambiguous, even contradictory, and often enough it could be used to criticize AEC policy.[64] The fact that policy makers in the AEC used this technical information selectively and sometimes misled the American public about the potential dangers of fallout should not obscure the important legitimizing role that scientists played in this process.

The AEC, supporting research on a grand scale, was apparently using scientific data in the decision-making process. By doing so, this controversial federal agency could cloak itself with the cultural authority of science. This was particularly true after the Eisenhower administration launched its Atoms for Peace program in 1953.[65] The program was an attempt to build public support for atomic energy in the wake of the first hydrogen bomb tests. It was also designed to put the Soviet Union on the defensive in terms of worldwide propaganda. One cornerstone of Atoms for Peace was support for basic research, specifically the use of radioisotopes as tracers. Tracer methodologies were sufficiently novel that the AEC isotope program attracted a great deal of interest from elite scientists, including ecologists. At the same time, the government used the program to publicize its commitment to the peaceful uses of atomic energy.

Mukerji presents a rather bleak picture of modern science. Scientists have exchanged independence and cultural legitimacy for financial support from the state. If so, it must be said that after some initial concerns and skepticism, few scientists showed much enthusiasm for returning to prewar patterns of research. But what were the motivations of ecosystem ecologists during the critical period of transition after World War II? Were they, to paraphrase a later critic, seduced by new economic opportunities?[66] Given our late twentieth-century biases cultivated by post–Vietnam War skepticism toward authority, the debacles at Three Mile Island and Chernobyl, and the recent revelations about safety violations at Hanford and Savannah River, there is a great irony in Eugene Odum, "Mr. Ecology," working so closely with the nuclear arms industry.[67] But his situation was hardly unique. Even the geneticist H. J. Muller, perhaps the most outspoken scientific critic of atomic energy policy, turned to the AEC for financial support. During the opening days of the Cold War few doubted the need for some nuclear deterrence. And behind this dark cloud there appeared to be a silver lining. For scientists and AEC administrators alike, nuclear technology promised a "scientific renaissance."[68] The delight of making new discoveries seemed impossible to resist. There was an almost universal optimism that the potential benefits of atomic energy outweighed its destructive power, that nuclear swords could eventually be beaten into plowshares. As AEC historians Richard Hewlett and Jack Holl conclude, "To bring that hope to reality was a strong and uplifting motivation."[69]

The ambiguities of the new relationship between science and the state were not entirely lost upon those who forged it. In his famous farewell address, an address in which the term "military-industrial complex" was coined, Dwight Eisenhower warned that the ideals of science might be perverted by the political process.[70] There was also the danger that science, as a partner in this new complex, might work to subvert democracy. Ecologists, too, sometimes ruminated on their new partnership with government. For Eugene Odum, nuclear technology was always a double-edged sword. Although confident that potential benefits outweighed environmental costs, he was forced to admit that a quarter century after Hiroshima, atomic energy remained primarily a military technology.[71]

7

The New Ecology

Ecologists can rally around the ecosystem as their basic unit just as molecular biologists now rally around the cell.

—EUGENE P. ODUM, "The New Ecology"

BY 1964 EUGENE ODUM was proclaiming the arrival of a "new ecology" that took the ecosystem as its fundamental object of study.[1] This new ecology was deeply rooted in historical tradition, but, as Odum acknowledged, its rapid rise owed much to atomic energy and other postwar developments. Odum and his younger brother Howard were at the center of this emerging specialty. Their award-winning research served as a model for early ecosystem studies. Together they pioneered the teaching of ecosystem ecology with advanced courses for college teachers at the Marine Biological Laboratory at Woods Hole (1957–1961). And they wrote the first textbook organized around the ecosystem concept. Eugene Odum's *Fundamentals of Ecology,* partly written by his brother, went through three editions (1953, 1959, 1971) and was translated into more than twenty languages.[2] No other textbook had such a profound influence upon the teaching of ecology during the 1960s; as one critic grumbled, for nearly twenty years the "odum" was the unit by which the success of ecology textbooks was measured.[3]

An Unusual Team

Eugene Odum was born in 1913, the son of the prominent sociologist Howard Washington Odum. The elder Odum was best known for his writings on American regionalism, the changes in southern society, and the effects of technology on social order. But he was also the scion of a strong tradition of organicism in American sociology. He apparently was an important influence upon his son's thinking, a man whose ideas Eugene would refer to again and again in his writings. Undoubtedly from his father Eugene inherited his commitment to organic holism, looking at the big picture in ecology.[4] This holistic approach was deepened and strengthened while he was a graduate student in zoology at the University of Illinois.

Odum entered the graduate school at a particularly exciting time. The chairman of the zoology department, Victor Shelford, was completing the manuscript of *Bio-Ecology*, a book coauthored with Frederic Clements. The book, though not a critical success, did provide the strongest statement of the organic point of view long championed by both authors. Odum's adviser, the young physiological ecologist Charles Kendeigh, also held some Clementsian ideas. However, the dissenting views of Henry Allan Gleason were also being taught, particularly by the plant ecologist Arthur Vestal. And Arthur Tansley's ecosystem paper appeared only two years before Odum arrived. Therefore, graduate students were exposed to a cross fire of ideas about the fundamental units of ecology.[5] Odum's dissertation, an experimental study of the heart rates of birds, may appear far removed from such theoretical concerns, but his interests in physiology and ecology meshed perfectly.[6] As Odum later recalled, "The transition from bird physiology to ecosystem function was quite natural for me since it involved moving up the hierarchy from physiological ecology of populations to the physiological ecology of ecosystems. It's really not such a big step to go from whole organism metabolism to community metabolism."[7] This physiological perspective became a hallmark of Odum's research.

The actual transition to ecosystem research came as a result of Odum's work for the Atomic Energy Commission, but it was stimulated by his brother who was completing a Ph.D. under G. Evelyn Hutchinson. During the late 1940s Howard was sending his older brother copies of Hutchinson's lecture notes, and Eugene himself was corresponding with the Yale limnologist.[8] Thus, *Fundamentals of Ecology* reflected a strong Hutchinsonian influence, particularly with respect to energetics and biogeochemical cycling. This influence was most noticeable in the chapter on energy that Howard contributed to the book, but looking back on this period Eugene also emphasized the degree to which his own early ideas on ecosystems were shaped by Hutchinson.[9]

Some of Eugene Odum's ideas have been controversial, but he has always remained within the broad mainstream of professional ecology. In contrast, Howard has cultivated an image as the enfant terrible of ecology, a specialist in unpopular ideas.[10] Controversy has followed him throughout his career. Nonplussed by a rather speculative introductory chapter in his dissertation on strontium cycling, Odum's graduate committee strongly recommended that Hutchinson remove the "philosophical vagaries" from his student's work. Hutchinson refused, and the dissertation retains its idiosyncratic fusion of biogeochemistry and evolutionary theory.[11] This episode, in 1950, foreshadowed what would become a somewhat stormy career. Controversy notwithstanding, Odum's image as an ecological outsider is a bit contrived. In attracting funding and graduate students he has had a remarkably successful career. Together with his older brother, he has been the recipient of prestigious scientific awards.[12] And his ideas have had a profound impact upon ecology, although sometimes indirectly through the writings of Eugene. But the fact that the older brother is perceived as a mainstream ecologist while the younger is perceived as a maverick reflects more than differences in style and personality. It also reflects subtle, but important, differences in the way the two men view nature. If Eugene has approached ecosystem ecology like a physiologist, then Howard has always approached nature like a physical scientist or engineer.

Like his brother, Howard emphasizes the important formative role that his father played in his intellectual development. His boyhood interests in biology were stimulated by Eugene, who was eleven years his senior, and by the ichthyologist R. E. Coker, for whom he worked after school. Unlike his brother, Howard was always strongly attracted to the physical sciences. As a youth he dabbled in electronics, a nascent interest that later found serious expression in his computer simulations of ecosystems.[13] During World War II he served as a meteorologist in the Air Force. As he later recalled, this experience with complex, large-scale natural phenomena taught him to appreciate the need to study systems holistically.[14] It also provided him with an entree to the biogeochemical research that he pursued as a graduate student.

Odum was introduced to Hutchinson through his father who was a visiting scholar at Yale after the war.[15] It was a propitious time to work with the limnologist, for his research program was approaching its zenith. Hutchinson, who was then participating in the Macy conferences on cybernetics, was attempting to complete a formal synthesis of biogeochemistry and population ecology. As a result, he was attracting an unusually diverse group of very bright graduate students. The grand synthesis never occurred, and his students all moved toward one pole or another; but during the late 1940s the intellectual environment around Hutchinson was extremely stimulating.

Odum attended one Macy conference with Hutchinson, an event that impressed him only by its confusion and lack of intellectual clarity; however, Hutchinson's paper, "Circular Causal Systems in Ecology," which was a product of the conferences, served as a model for Odum's early work. Even more influential was Alfred Lotka's, *Elements of Physical Biology*, a book that Hutchinson recommended to his student.[16] Published in 1925, Lotka's book attempted to define a new area of biological research: physical biology. Physical biology was to be the application of physical principles to complex biological systems, particularly that all-encompassing system that we now refer to as the biosphere. From Lotka's perspective, that of the physical chemist, all biological processes could be reduced to exchanges of matter and energy among the compartments of a system. As such, biological systems were governed by the laws of thermodynamics. But biological systems differed in two important ways from the type of closed chemical systems traditionally considered in thermodynamics. They were more complex, and, unlike isolated chemical systems, they were open to continuous inputs and losses of energy. Therefore, biological systems never came to a true equilibrium state, defined in terms of maximum entropy, but rather attained a steady state, defined in terms of constant energy flow. Lotka's ideas on energy flow were not widely adopted by ecologists during the 1920s and 1930s, but particularly through the writings of Howard Odum they had a pervasive influence upon the way later ecologists thought about ecosystems.

The ideas of open systems and steady states could be applied very broadly in biology. Energy transfer associated with predation and the cycling of elements were obvious cases. But Lotka also used his open systems approach to discuss population dynamics in kinetic terms, where the size of the population was a function of the constant entry and removal of individuals. This general approach was later expanded upon by Hutchinson and some of his students. Less propitiously, Lotka also attempted to explain evolution in thermodynamic terms. Evolution, for Lotka, was not so much species changing over time, but rather the overall accumulation and distribution of energy within a system. Natural selection always maximized the flow of energy and matter through this system. This curious idea seems to have been almost completely ignored by more traditional evolutionary biologists. Referring to it as the "maximum power principle," Odum made it the leitmotif of his controversial evolutionary writings.

By the time that Odum was a graduate student, steady state thermodynamics was attracting considerable attention in scientific circles.[17]

Hutchinson encouraged his students to explore this literature, and the ideas, particularly Lotka's, were widely discussed among the Yale ecologists. Given his background in the physical sciences, particularly his training in meteorology and physical chemistry, it is perhaps not surprising that Odum should have been so attracted to Lotka's book. As Peter Taylor has pointed out in a perceptive article, Lotka had a much greater influence upon Odum than he did on more biologically oriented ecologists.[18] Whereas Hutchinson and others only borrowed Lotka's mathematical models, Odum fully grasped the intent of Lotka's physical biology. More than almost any other ecologist, he has approached the study of ecosystems as a physical scientist might. As we shall see, this approach often brought him into conflict with other biologists, particularly more traditional evolutionary ecologists.

Within months after defending his dissertation, Howard Odum was writing the chapter on energetics for his brother's textbook. Together the two made an unusual team. In terms of personality, scientific style, and intellectual perspective they were very different, and during the years when they were establishing the "new ecology" there was a bit of sibling rivalry. In important ways, however, their talents complemented one another. The success of *Fundamentals of Ecology* reflects this fruitful collaboration. One can trace much of the evolution of the ecosystem concept through the subsequent editions of the textbook. It not only presented the ideas of the two Odums as they developed over two decades, but the textbook also provided a synthesis of ideas from other leading ecosystem ecologists. One of Eugene Odum's most important skills has been his ability to make often abstruse theory intelligible to a general audience. As a result, what might have remained part of a rather narrow, technical literature was brought into the mainstream of biological thought. Among ecosystem ecologists he may not have been the most original thinker, but through his semi-popular writings Eugene Odum was easily the most influential. For admirers and critics alike, the name Odum became indelibly linked with ecosystem ecology.

A Refined Concept

It is useful to think of scientific concepts as modular constructs. The parts fit together to form a more or less unified whole, but some parts can be removed or replaced without destroying it. As a concept becomes refined, original ideas may even be replaced by their contradictories.[19] Such is the case with the ecosystem concept as it evolved during the 1950s and early 1960s. Nearly all ecologists accepted some key components, although the scientists might interpret them slightly differently. Some original ideas were abandoned or replaced by quite different ones. Other components remained unsettled. One could conceivably reject one or more of these and still be an ecosystem ecologist. What follows is a kind of conceptual map of these ideas.

The Machinery of Nature

Two of Tansley's conceptual innovations remained at the core of ecosystem ecology. Unlike population and community, which were exclusively biological, the ecosystem was the one commonly used ecological concept that encompassed both biotic and abiotic factors. As such, it was a particularly useful term for discussing biogeochemical cycles and energy flow. It also became useful for discussing many practical environmental problems that could not be clearly demarcated along biological and nonbiological lines. Rival terms, such as biogeocoenosis, never gained much of a following outside the Soviet Union. Tansley's notion that systems are defined by the investigator also proved to be a durable innovation. Ecologists had long bickered about the natural boundaries of plant and animal communities. For most ecosystem ecologists, such arguments were both fruitless and irrelevant. The ecosystem was not so much a concrete geographical entity as a flexible abstraction. An ecosystem such as a small pond was not completely isolated from its surroundings. Although its boundaries might be poorly marked, the investigator could still define it as a "system" for the purpose of ecological study. Energy, chemical substances, or organisms might regularly move in and out of the system, but these movements could be treated as simple gains and losses in the ecosystem budget. The flexibility that this allowed in pursuing studies of energy flow and material cycling was very attractive, and this original idea was also widely accepted by later ecologists.

Some of Tansley's other ideas changed, but in subtle ways. The term *ecosystem* had arisen out of Tansley's critique of Clementsian ecology. Despite his fascination with the philosophy of modern physics, neither Tansley nor later ecosystem ecologists, ever completely abandoned Clements's organismal metaphor, although after World War II, there was what Peter Taylor describes as a "partial transformation" of ecological metaphors, a gradual shift from organic to machine images.[20] One might, for example, contrast the organismal allusions in Raymond Lindeman's trophic-dynamic paper with

George Clarke's postwar picture of the ecosystem as a set of interlocking gears.[21] Of course, the notion that something can be both organism and machine is not unusual in biology. And the postwar development of cybernetics, in which ecology took part, encouraged analogies between the industrial and organic worlds.

The Odums's writings exemplify this januslike conception of ecosystems. Eugene Odum, trained within a basically Clementsian tradition, has always stressed the organismal attributes of ecosystems: development, metabolism, and homeostasis. Ecosystems are not, in fact, organisms, although one can draw useful parallels between them.[22] For example, when he needs an analogy for communication and control mechanisms in the ecosystem, Eugene turns to the physiology of the endocrine system.[23] For Howard Odum, ecosystems are only very complex machines. Processes in ecosystems are fundamentally no different from those of water wheels, steam engines, and electrical circuits. Writing in 1959, he stated, "The relationships between producer plants and consumer animals, between predator and prey, not to mention the numbers and kinds of organisms in a given environment, are all limited and controlled by the same basic laws which govern non-living systems, such as electric motors and automobiles."[24] When Howard Odum speaks of communication and control, he rarely turns to organic analogies; instead, he speaks of the invisible wires in nature's circuits.[25]

The complementarity of these images is most evident in the Odums's writings on self-regulation and the steady state. Ideas of stability and self-regulation permeate the writings of both Odums, going back to their Ph.D. dissertations. Beginning with the second edition of *Fundamentals of Ecology* (1959), Eugene Odum began discussing these phenomena explicitly in terms of homeostasis.[26] He had read Walter B. Cannon's *The Wisdom of the Body* as a graduate student during the late 1930s, and he later used it as required text in the physiology course that he taught at the University of Georgia.[27] Cannon's concept of homeostasis—the idea that organisms are capable of maintaining internal stability in a fluctuating environment—was directly applicable to Odum's early research in physiological ecology. But Cannon's approach also held more general appeal for Odum who shared the Harvard physiologist's commitment to functionalism, the holistic study of complex biological systems, and the close analogy between biological and social entities.[28]

For Odum, homeostasis became a general biological principle. Homeostatic mechanisms acted at all levels of biological organization from cells to ecosystems. Thus, Odum was not simply reformulating the Clementsian argument that the ecosystem is a kind of organism because it is homeostatic. Rather he was making the much stronger claim that all living systems—cells, organisms, populations and ecosystems—share this common self-regulatory property. Part and parcel of this way of looking at the living world is the acceptance of biological functionalism. For Odum, individual organisms and populations really are *parts* of the ecosystem in the sense that they carry out particular functional roles. Individuals might compete with one another, but if an ecosystem is to survive, competition must be balanced by cooperation. Thus, the evolution of stability in ecosystems occurs through a reduction in competition and an increase in mutualism. For example, Odum would like to see the evolution of lichens as a progressive change—from the primitive condition where a fungus parasitizes its algal host to a truly mutualistic association where both fungus and algae benefit. "Like a balanced equation," Odum wrote, "it seems reasonable to assume that negative and positive relations between populations eventually tend to balance one another if the ecosystem is to achieve any kind of stability."[29]

Odum believed that the evolution of homeostasis and the consequent stability of ecosystems occurred through a combination of group selection and coevolution. By the late 1960s such statements were highly controversial, but during the 1950s they reflected mainstream biological thought. As Gregg Mitman has described in his detailed study of ecology at the University of Chicago, Alfred Emerson had been expressing similar ideas about homeostasis and evolution for more than a decade.[30] Odum was familiar with Emerson's work on social insects and may have derived some of his ideas from it. But the concept of homeostasis was being used so widely after World War II that tracing lines of influence is difficult. During the two decades following the war, a diverse group of scientists used homeostasis to explain a broad range of biological phenomena.[31]

In contrast to his brother's more biological account of ecosystem stability, Howard Odum attempted to provide a thermodynamic explanation. Citing Lotka's *Elements of Physical Biology,* he argued that natural selection favored those systems that maximized power output in the form of growth, reproduction, and maintenance. But maximum power required a sacrifice in efficiency. Using the example of Atwood's machine, a simple system composed of two weights connected by a rope attached to a pulley, Odum noted that maximum power output is attained at 50 percent efficiency. In short, the optimum efficiency for power production was much less than maximum efficiency. This "maximum power principle," Odum believed, applied not only to simple laboratory demonstrations but to all systems, including the complex open systems of the biological world. Power was at a premium in the struggle for existence, but it was not the only factor for survival. Natural selection favored not only the most powerful systems but also the most stable. This "stability principle," which Odum attributed variously to Lotka and the University of California zoologist Samuel J. Holmes, stated that as living systems evolve, increasing amounts of energy are diverted to maintenance.[32] In a climax ecosystem, like the coral reef at Eniwetok Atoll that he and his brother had studied, nearly all the energy trapped by the producers was used to maintain the complex living community. "As an open system," he wrote of the reef, "the construction of self-regulating interactions has led by selective process to the survival of the stable."[33] Such stable ecosystems were in a thermodynamic steady state; photosynthesis was almost completely balanced by respiration.

The maximum power principle and the stability principle could be easily translated into the language of homeostasis. Walter B. Cannon made an important observation in *The Wisdom of the Body:* organisms often sacrifice efficiency to maintain stability. Large amounts of energy are used to maintain the proper balance of water and minerals in the body.[34] Although this might appear uneconomical, Cannon believed that homeostatic stability was a necessary precondition for the evolution of the vertebrates. This fusion of physiological and physical metaphors for ecosystem stability became even tighter during the 1960s when the language of cybernetics began to enter ecology. Cybernetics, as originally conceived by Norbert Wiener, had drawn heavily upon Cannon's physiology, but after the war many biologists used cybernetics as a new way to discuss organic self-regulation. For Howard Odum, feedback loops in ecosystems could be diagrammed as if they were parts of an electronic circuit. For Eugene Odum, these loops were more closely analogous to hormonal or neural control systems. But as his later writings suggest, homeostasis and cybernetics were simply two different ways of discussing the same thing.[35]

Systems Thinking and Systems Ecology

Cybernetics was just one example of a broader intellectual movement that developed after World War II, a development that Robert Lilienfeld has referred to as "systems thinking."[36] New areas of research such as cybernetics, general systems theory, operations research, game theory, information theory, and computer simulation were premised upon the belief that diverse physical, biological, and social entities could be treated as systems. In his highly critical history, Lilienfeld identifies two related characteristics with the rise of systems thinking. One is *syncretism*, the tendency for what were once distinct technical disciplines to fuse. Many supporters of systems thinking advocated interdisciplinary sharing; indeed, they often saw systems theories as tools for breaking down disciplinary barriers. The other characteristic is *migration*, the movement of systems thinking into previously well-established disciplines. This migration can take the form of technical innovation, but, according to Lilienfeld, it is also characterized by missionary essays, programmatic statements, and the rhetorical use of technical jargon.[37] Systems Ecology, a term coined by Eugene Odum, partook of both syncretism and migration.[38]

Historians have emphasized the difficulty of adequately defining systems ecology.[39] This might be expected in a case where several scientists more or less independently borrowed from an already heterogeneous mix of ideas; however, the situation is not entirely chaotic. For the purposes of this discussion, I identify three broadly overlapping definitions of systems ecology. In a very general sense, ecosystem ecology and systems ecology could be used as synonyms. Systems ecology could be much more narrowly identified with the small, highly technical subdiscipline of ecosystem modeling. Finally, one might identify systems ecology with Howard Odum's ambitious and idiosyncratic attempt to create a universal science of systems.

The general use of the term is best exemplified by the work of Eugene Odum. For him, systems ecology was as much a "state of mind" as it was a set of mathematical techniques.[40] It involved thinking about the living world in the way Arthur Tansley had suggested during the 1930s. The philosophical core of Tansley's ecosystem concept was the belief that nature is composed of innumerable, partially overlapping systems. These spatiotemporal units could be of any size; an ecosystem might be an aquarium, a space capsule, a farmer's field, a pond, or the entire biosphere.[41] Thus, to be an ecosystem ecologist required one to think in terms of systems; as Odum succinctly put it, "the new ecology is thus a systems ecology."[42] Odum often borrowed from the language of cybernetics and other systems sciences to discuss ecological phenomena. For example, the idea of negative feedback control could be used interchangeably with the more biological idea of homeostasis. But this was borrowing, or, to use Lilienfeld's metaphor, it represented the migration of cybernetic ideas into the established discipline of ecology. There is little evidence that Odum himself delved deeply into the technical literature of cybernetics or systems theory.

Significantly, when it came time to prepare the third edition of *Fundamentals of Ecology,* the new chapter on systems ecology was written not by Odum, but by the younger biologist Carl J. Walters.

If systems ecology could be used synonymously with ecosystem ecology, then it could also be identified with a much smaller subspecialty that began to emerge during the early 1960s. For ecologists such as Ramon Margalef, Howard Odum, Jerry Olson, Bernard Patten, Lawrence Slobodkin, George Van Dyne, and Kenneth Watt, systems sciences offered an array of new tools for dealing with complexity. "The significant features of the approaches," Patten wrote, "are that they are essentially mathematical in nature, and they require modern computing equipment for effective application."[43] For Patten and others, the ecosystem might be a "keystone" concept, but it was one that could be explained fully only in the language of mathematics. This smaller group of "systems ecologists" was distinguished by two important characteristics. In contrast to Eugene Odum, for whom systems thinking was primarily qualitative, this group more fully exploited the technical innovations of the systems sciences. And rather than simply borrowing systems thinking, these ecologists contributed to the journals and interdisciplinary conferences that had sprung up around the new systems sciences. They were enthusiastic proponents of what Lilienfeld refers to as the syncretism of systems thinking.

If systems ecology involved the migration of new ideas into an established discipline, this was not accompanied by a migration of scientists. Almost without exception, the systems ecologists were trained in traditional biological programs.[44] Typical of this group was Bernard Patten. As an undergraduate zoology major at Cornell, Patten took no courses in mathematics. He began working on a master's degree in botany at Rutgers University in 1954 but was inducted into the Army. While doing pesticide research at Fort Detrick, Maryland, he was introduced to a series of essays on the use of information theory in biology. He later recalled that his inability to follow the mathematics made the essays all the more intriguing.[45] Once he had "caught the systems bug," Patten read widely in the literature of the systems sciences, and he took several courses in mathematics as he completed his Ph.D. His dissertation used information theory to measure the diversity of a phytoplankton community.

As a young professor at the College of William and Mary, Patten assigned Ross Ashby's *Introduction to Cybernetics* as a textbook in his marine ecology course. Taught during the early 1960s, this may have been the first course in "systems ecology." But the real development of the subspecialty came when Patten moved to the Oak Ridge National Laboratory in 1963. The laboratory, with its heavy emphasis on the physical sciences, had the advanced computers needed for sophisticated modeling, and two other biologists on the staff shared Patten's enthusiasm for systems ecology: Jerry Olson and George Van Dyne.

Like Patten, both Olson and Van Dyne came from traditional biological programs. Olson received a Ph.D. in botany from the University of Chicago in 1951 with a dissertation on soil changes during dune succession. Often compared favorably to Cowles's classic study of the same dunes, Olson's doctoral research won the prestigious Mercer Award from the Ecological Society of America. Van Dyne was the product of an agricultural program. He had a bachelor's degree in agriculture from Colorado A & M College and a master's in animal husbandry from South Dakota State University. His Ph.D. was in nutrition from the University of California at Davis with a dissertation on the effects of cattle grazing on grassland ecosystems. Although interested in ecological modeling, Van Dyne had little experience with computers before he arrived at Oak Ridge.

Olson was the first of the three to arrive at Oak Ridge. By 1960 he was continuing his soil studies with the use of radioactive tracers and experimenting with the use of analog computers to simulate ecosystems. He was perhaps the first ecologist to do so. Patten arrived at Oak Ridge in 1963, followed a year later by Van Dyne. The interests and abilities of the three ecologists apparently complemented one another. Patten later recalled that his experience at Oak Ridge was exciting, even exuberant.[46] Funded by grants from the Ford Foundation and the National Science Foundation, the group offered an advanced course in systems ecology through the nearby University of Tennessee. Although it was intended primarily for graduate students, the course was audited by a number of postdoctoral fellows, visiting scientists, and faculty members.[47] This fruitful collaboration was short lived. By 1968 both Patten and Van Dyne had moved back to academia. But the team had left its mark: after 1968 Oak Ridge was recognized as the leading center for systems ecology in the United States.

Systems ecology was as diverse as the systems sciences that it drew upon, but at the core of this enterprise was the use of computers for modeling and simulation. Increasingly, the new digital computers were used for this purpose. Van Dyne, who had learned FORTRAN programming, had already begun to explore its possibilities during his brief stay at Oak Ridge. But the early development of systems ecology relied more heavily upon analog computers. In an analog computer electrical circuits are used to represent the features of the system under study. More to the point, the circuits represent mathematical equations describing these features. During a simulation, the computer generates voltages that behave like the mathematical variables in the equations. In this way several equations can be solved simultaneously. For the systems ecologist, analog computers served two important purposes.[48] They could be used to model very large, complex systems. Physical and biological processes occurring over the course of several decades could be rather quickly simulated in the laboratory. This had tremendous theoretical and practical implications for predicting the behavior of real ecosystems. But the very process of building analog models also played an important heuristic function. Modeling involved a stepwise process of abstraction from the complexity of nature to the relative simplicity of the analog circuitry. For example, the ecologist might identify and isolate food chains within an ecosystem, draw compartment diagrams representing the flow of energy through these chains, write mathematical equations representing the rates of energy flow, and finally create the analog program (i.e., circuit diagram).[49] This process of abstraction could be a powerful stimulant for the scientific imagination, leading to new questions about the real system under study. In a more general way, systems ecology could also serve as a "strategy of research," a set of procedures that forced the ecologist to keep the "big picture" in mind.[50] By building models explicitly in terms of systems and component subsystems, one might hope to achieve a coherent explanation, rather than a collection of fragmentary results.

There is no reason why the type of systems ecology described above should necessarily be restricted to the study of ecosystems, and, in fact, ecologists such as Kenneth Watt and C. S. Holling used systems analysis to study populations. Thus, there is some justification for claiming that systems analysis has become a standard tool in ecology.[51] However, for better or worse, systems ecology and ecosystem ecology became closely linked. When supporters, and especially critics, spoke of systems ecology, they were usually speaking within the context of ecosystem studies. Partly this was a result of Eugene Odum's "missionary essays" that equated the two. Partly it was owing to the boost that both ecosystem ecology and systems ecology received from the International Biological Program (discussed in chapter 9). Partly, it was also responding to the way that Howard Odum used both the ecosystem concept and systems ecology in his attempt to build a grand systems science.

Independent of the Oak Ridge group, Howard Odum also experimented with analog models of the ecosystem during the early 1960s. He and his students published a number of important modeling papers, including two in the high visibility journals *Science* and the *American Scientist.*[52] But he was hampered by a lack of sophisticated computer equipment, and thus he turned to a more conceptual approach to systems ecology. In early analog models, ecological variables were simulated by actual electronic components such as resistors and capacitors; however, by the end of the 1960s these were replaced in Odum's circuit models by a more general set of energy symbols. From Odum's perspective this new approach had important advantages.[53] It freed him from some physical restrictions imposed by electrical circuits, and it allowed him to develop a universal energy language (*energese*) for modeling systems, in general. Odum believed that this language could be applied to any system: electrical, mechanical, biological, or social. This ambitious program in systems ecology was summarized in Odum's semipopular book, *Environment, Power, and Society* (1971).

Odum's book, an eclectic and idiosyncratic work aimed at a general audience, was intended to explain basic concepts in ecology using Odum's energy language.[54] The early chapters provided an excellent summary of ecosystem ecology and a coherent introduction to systems modeling. It also presented a cogent argument for the limits of industrial growth. Circuit diagrams were skillfully used to illustrate the dependence of agricultural ecosystems and industrial societies upon fossil fuel subsidies. Left at that, the book might have become one of the most important environmental treatises of the 1970s. Less auspiciously, in later chapters Odum applied his systems approach to politics and religion. "The energetic laws are as much first principles of political science as they are first principles of any other process on earth," he claimed.[55] Voting, public opinion, taxes, even revolution and war could be expressed in the language of energy circuits. This large-scale reduction of complex social phenomena to simple quantitative variables exemplified one of the "besetting vices" that Lilienfeld has identified with systems thinking.[56] Lilienfeld complains that by mechanizing human behavior and social interactions, systems thinking inevitably leads to authoritarianism. Odum's book is a case in point. Although he claimed to be defending democracy, the simple control loops of his energy diagrams were more suggestive of coercion and manipulation than of individual freedom.

Odum's excursion into social and political discourse was unfortunate. *Environment, Power and Society,* with its penetrating insights into environmental problems, might have been influential, but apparently it was not taken seriously by professional ecologists. Book reviewers in many leading journals read by ecologists simply ignored the book. The review in *Science,* although not entirely hostile, emphasized the idiosyncratic character of Odum's systems thinking. Egbert Leigh, another student of G. Evelyn Hutchinson, praised Odum for undertaking such an ambitious project, but then he added that the book was "a most maddening work, which at first sight seems totally undisciplined, a chaotic mixture of the asinine, the banal, and the brilliant, with random observations, often in conflict with the available evidence, on nearly everything under the sun."[57] Unfortunately, Odum seemed to be repeating the fate that had befallen his intellectual model, Alfred Lotka, a scientist who never successfully reached his intended audience. Like Lotka, much of Odum's influence in ecology, particularly in later years, came indirectly through the writings of others. The reception of *Environment, Power, and Society* was a clear signal that by the early 1970s, he had moved to the fringes of systems ecology and professional ecology in general.

Dealing with Complexity

Arthur Tansley first presented the ecosystem concept within the context of a critique of holism. After more than half a century his trenchant analysis of this philosophical position has lost none of its original bite. Ironically, as it evolved, the ecosystem concept became closely identified with the very philosophy that Tansley so adamantly opposed. This change came about primarily through the writings of Eugene Odum and Howard Odum, strong defenders of holism. Antireductionistic themes increasingly permeated *Fundamentals of Ecology* as it went through three editions. Several other ecosystem ecologists, particularly those drawn to modeling and systems analysis, also claimed to be using holistic approaches in their work. As a result, both adherents and opponents of the specialty have tended to identify ecosystem research with philosophical holism.

Biological debates over holism and reductionism almost always generate more smoke than light, and those in ecology are no exception. As a few perceptive critics have pointed out, Howard Odum's attempt to explain all ecological phenomena in terms of energy looked suspiciously like reductionism, albeit "large-scale reductionism."[58] Conversely, it can be argued that self-styled ecological reductionists have rarely tried to really explain ecosystem phenomena in terms of smaller biological units.[59] They simply ignore the "whole" and study the "parts" in isolation. Such diffuse and polemical debates over reductionism are not uncommon in the history of biology, and more recently some ecologists have been quite sophisticated in explaining the hierarchical relationships among biological systems.[60] But the historical question remains: What purpose did holism play in the early development of the ecosystem concept?

The answer to this question is twofold: holism provided useful rhetorical arguments for justifying and legitimizing an emerging specialty, and it served as a fruitful heuristic for stimulating research. Holistic arguments rarely converted skeptics, but they did increase solidarity within a small group of practitioners. When Eugene Odum arrived at the University of Georgia he encountered a zoology department that was largely indifferent toward ecology. For most of his colleagues, ecology was simply another name for natural history. The idea of ecosystem research was almost totally unknown at the university. Therefore, Odum had to justify not only studying large biological systems but also studying them in terms of their overall metabolism. Moreover, outside this local context where he was attempting to establish an independent institute of ecology, Odum needed to justify the autonomy of ecosystem ecology within the professional discipline of ecology where he was trying to establish a new specialty and within the still broader context of the scientific community where he was competing for funding. In *Fundamentals of Ecology,* Odum presented an unusual and sophisticated defense.

The first part of the argument was antireductionistic. Although nature is organized on many levels, he argued, no level is necessarily more complex or difficult to study than any other.

> *When we consider the unique characteristics which develop at each level,* there is no reason to suppose that any level is any more difficult or any easier to study quantitatively. The enumeration and study of the units of an organism (i.e., the cells and tissues) is not inherently any easier or more difficult than the enumeration and study of the units of a community (i.e., the organisms). Likewise, growth and metabolism may be effectively studied at the cellular level or the ecosystem level by using units of measurement of a different order of magnitude.[61]

To make this point more explicit, Odum drew a diagram of the units of biological organization from cells to the biosphere, but the diagram was arranged horizontally, rather than vertically. The message was clear: if all levels are equally complex, then there is no reason for biologists to adopt a reductionist strategy. Research at the level of ecosystems is just as likely to produce important discoveries as research in molecular biology.

The second part of Odum's argument was more holistic. Dogmatic reductionism impeded the advance of science. If biologists spent all their time studying cells, then they would never get around to studying populations and ecosystems. Furthermore, although all levels of biological organization shared common characteristics (growth, development, metabolism, and homeostasis), the mechanisms by which these processes occurred were different. Understanding a process at one level only partly explained the same process at another level. Knowing about homeostasis at the cellular or organismal levels might provide ecological insights, but it could never completely explain the stability of populations and ecosystems. Thus, science had to advance along a broad front. "This situation is analogous to the advance of an army." Odum concluded, "A breakthrough may occur anywhere, and when one does, the thrust will not penetrate far until the whole front moves up."[62]

Systems thinking often goes hand in hand with philosophical holism. Not surprisingly, the systems ecologists were among its most ardent supporters. Using one of his clever metaphors, Howard Odum referred to the "macroscope" of systems science.[63] In contrast to the microscope that allowed the scientist to observe hidden details, the macroscope served as a kind of "detail eliminator." Through the macroscope large systems appeared simpler; they became black boxes with inputs and outputs. Freed from detail, the systems ecologist could ask new questions about the behavior of the system as a whole. The intricate biological details of a particular ecosystem were relevant; natural history served as an important means of creating an "inventory of parts" for the system, but the real explanation came in terms of overall energy flow through the ecosystem as a whole.[64] For example, when the Odums had studied the metabolism of the reef at Eniwetok Atoll, they were not concerned with individual species. Indeed, at the time they were unable to identify them. Nonetheless, they were able to estimate the total flow of energy through the entire system. Had they started studying the reef from the bottom up, they might never have gotten around to studying its overall metabolism.

A Managerial Ethos

Environmentalists of the 1970s frequently used the ecosystem concept to argue for the preservation of natural habitats. Ecosystem ecologists were also deeply committed to conservation, but not necessarily in the form advocated by popular environmental groups. In his book, *Nature's Economy,* Donald Worster argues that professional ecologists are imbued with a "managerial ethos," a set of beliefs that reflect an imperialistic attitude toward nature.[65] Whether or not one completely accepts Worster's critique, the ecosystem concept did become closely identified with the rational management of nature for human benefit. In the first edition of his textbook, Eugene Odum wrote, "The aim of good conservation is to insure a continuous yield of useful plants, animals, and materials, by establishing a balanced cycle of harvest and renewal. . . . The principle of the ecosystem, therefore, is the basic and most important principle underlying conservation."[66] Ecology provided an understanding of natural cycles and the homeostatic limits of ecosystems. Humans could learn valuable lessons from nature, but this knowledge also allowed humans to intervene, to manipulate natural ecosystems, and to create artificial ones.

The idea of "man the manipulator" is a common thread running through the literature of ecosystem ecology.[67] For many ecologists this phrase encapsulated the most pressing dilemma facing science and society. Technology, as Odum acknowledged, was a double-edged sword.[68] It not only held out the promise of a better life, but it also held the potential for destroying the environment. Faced with this dilemma, ecosystem ecologists rarely turned away from technology. Instead, they often looked forward to a new era of "ecological engineering."[69] The ecological engineer used nature's own machinery to construct new life support systems for human society. For example, Howard Odum envisioned a situation where municipal waste water might be pumped through a seminatural, aquatic ecosystem. Nutrients would be removed by microbes that would serve as a base for various food chains. At the top of the food chains might be harvestable species such as crabs. Clean effluent leaving the ecosystem could be reused as drinking water by the human population. Thus, by combining human engineering principles with nature's own "self-design principles" a true "partnership with nature" might be achieved. Much to the dismay of some traditional environmentalists, such plans were actually implemented on a small scale, and they apparently worked quite well.[70]

Peter Taylor has characterized Howard Odum's attitude toward ecological engineering as "technocratic optimism."[71] However, this technocratic optimism rested uneasily with a more apocalyptic sense that the real cutting edge of technology's sword was its destructive, rather than its productive, edge. Modern industrial society with its voracious appetite for energy and its ability to alter natural ecosystems had precipitated an environmental crisis. In a section of *Environment, Power, and Society,* "The Network Nightmare," Odum used the fantasy of a monstrous computer-organism gone berserk as a metaphor for this crisis.[72] Although he discussed the potential for ecological engineering, Odum suggested that completely managing nature was an impossibility. His brother was also uneasy about the managerial ethos. Homeostasis had evolved gradually in ecosystems, and human perturbations often destroyed the intricate self-regulatory mechanisms of nature. Human domination of nature was a "dangerous philosophy," a belief that Eugene Odum wished to dispel with his *Fundamentals of Ecology.* "When the reader has finished with this book," he wrote, "I am sure he will agree that we cannot safely take over the management of everything!"[73]

Extending the Research Agenda

Raymond Lindeman's trophic-dynamic paper identified three related intellectual problems that together provided the primary foci for ecosystem ecology: biogeochemical cycling, energy flow, and succession. These problems held an intrinsic theoretical interest for those interested in how ecosystems functioned. But for reasons that Lindeman could not have imagined these problems also loomed large in public debates after World War II. Less than a year before Hiroshima, Vladimir Vernadsky wrote that for the first time in history man was becoming "a large-scale geological force."[74] Human activities were now capable of altering not only local environments but also the biosphere as a whole. No example illustrates the convergence of theoretical and practical uses of ecological knowledge so clearly as Howard Odum's dissertation on the biogeochemistry of strontium. When Odum began his research strontium was a substance practically unknown to the nonscientific public. During the late 1940s when Odum began his research, studying strontium was simply another step in G. Evelyn Hutchinson's ambitious biogeochemical survey of the elements. Biogeochemically, strontium acted in a qualitatively similar way to the more abundant and biologically important element calcium. However, the point of Odum's research was to demonstrate that the strontium cycle, or what he referred to as the "strontium ecosystem," was a stable, self-regulating cycle.[75] According to Odum as the concentration of strontium increased in the oceans it was removed and deposited in the exoskeletons of molluscs. As a result of this simple control mechanism strontium was maintained in a steady state equilibrium. He provided data showing that levels of strontium in the oceans had not changed significantly during the past 600 million years.

Odum's early research was the epitome of basic science, and there was no hint that the strontium cycle was of more than purely academic interest. A decade later, atmospheric testing of nuclear weapons and the consequent release of large amounts of radioactive strontium-90 made this element familiar to almost every educated American.[76] By the late 1960s George Woodwell and others demonstrated that toxic substances such as DDT also had biogeochemical cycles.[77] On a local level, these substances could become concentrated in the higher levels of food chains. Global cycles distributed some of these persistent chemicals far from their points of origin. Even Antarctica was not free of pesticide residues. For a public already conditioned by Rachel Carson's *Silent Spring,* such widely publicized revelations heightened environmental concerns. For professional ecosystem ecologists, the discoveries simply confirmed what they already believed: biogeochemical cycles were fundamental processes in all ecosystems.

Biogeochemical cycles were important, but for most ecosystem ecologists energy flow was of even greater interest. As one critic later complained, the specialty seemed to be "obsessed with calories."[78] As the ultimate limiting factor for life, energy flow seemed to hold the key for understanding the structure and function of ecosystems. Expressing the appeal of this line of research, John Teal wrote, "The study of community metabolism is one means of making a functional analysis of an ecosystem. . . . It provides a measure of the total activity of the community just as a study of individual metabolism does for an individual organism."[79] In theory, the structure and function of the living community—the sizes, numbers, and kinds of organisms; the relationship between producers and consumers; competitive interactions among species; the interdependence of predators and prey—could all be explained in terms of energy transformations. As G. Evelyn Hutchinson pointed out in his tribute to the young ecologist, Raymond Lindeman had taken the first tentative steps toward reducing the complexity of ecosystems to such simple energetic terms.[80] Hutchinson's later student, Howard Odum, was primarily responsible for continuing this intellectual process. As Peter Taylor points out, Odum was unique among ecosystem ecologists in that he reduced all ecological parameters—biomass, population sizes, diversity, essential chemical elements—into energy.[81] Following Lotka, he intended to explain ecological phenomena completely in terms of the principles of thermodynamics. Although his grand synthesis, *Environment, Power, and Society,* apparently had little impact on professional ecologists, some of his earlier conceptual innovations were highly influential.

More than any other ecologist, Howard Odum has shaped the way biologists think about energy. The chapters on energetics that he wrote for his brother's textbooks summarized several of his important technical papers and presented this information to a broad audience. Biologists who never read his massive study of Silver Springs, nevertheless absorbed its message through textbook accounts. Perhaps the major conceptual innovation to come out of this study was Odum's pictorial model of energy flow. Unlike earlier diagrams, which confused the movement of energy and material, Odum's model showed at a glance what was important about energy flow. Each trophic level was represented by a rectangular compartment, the size of which represented its energy content. The compartments were connected by arrows representing the flow of energy from one trophic level to the next. Arrows representing the heat loss of respiration were also drawn from each compartment. Diagrammatically the arrows illustrated the effects of the second law of thermodynamics; every trophic transfer entails a loss of usable energy from the system. Like the orbital or "electron cloud" diagrams used by chemists to visualize the probable locations of electrons around the nucleus, Odum's pictorial model played an important explanatory role in ecosystem energetics. Once seen, it became difficult to think about energy flow removed from the visual context of the diagram (figure 8). Not surprisingly, this model appears today in virtually every undergraduate textbook of ecology. Ironically, Odum's later electrical circuit diagrams, which he considered superior to the compartment models, never caught on among ecologists. Pictures are important, and apparently only the right picture can capture the essence of a complex natural process.

Cracks in the Edifice

By the mid-1960s the Odums were at the height of their influence. During this period Eugene Odum wrote perhaps his most important and controversial article, "The Strategy of Ecosystem Development."[82] Published at the end of the decade, the paper was based on Odum's 1966 presidential address to the Ecological Society of America. The apotheosis of the "new ecology," it summarized two decades of research on energy flow and biogeochemistry and placed it within the conceptual framework that the Odum brothers had constructed. Unlike many such articles, however, this was not simply another programmatic statement. In a table, Odum set out a list of twenty-four universal trends that he claimed were characteristic of succession. One might quibble whether these constituted hypotheses and predictions in the technical sense. But the sharp dichotomies that he drew between immature and mature ecosystems were striking, and some ecologists treated them as predictions to be empirically tested.[83]

Strategy meant maximizing stability, a process that occurred on a number of time scales. "In a word," Odum wrote, "the 'strategy' of succession as a short-term process is basically the same as the 'strategy' of long-term evolutionary development of the biosphere—namely, increased control of, or homeostasis with, the physical environment in the sense of achieving maximum protection from its perturbations."[84] If ecosystems actually employed this strategy, a number of measurable changes ought to be observed. During succession diversity should increase as species became specialized for particular functional roles. Therefore, the number of species ought to be greatest in mature, climax ecosystems. Specialization promoted symbiosis as species became more dependent upon one another. In the language of systems ecology, this increase in structural complexity meant an increase in information content and a decrease in entropy in the system as a whole. The growth of decomposer populations and detritus-based food chains would increase the efficiency of nutrient cycling. As a result the leakage of nutrients from the system would decrease. During succession the amount of biomass would increase until it reached a maximum at climax; however, the rate of production in relation to the rate of respiration would decrease. In the climax ecosystem, gross production would roughly equal respiration. In other words, virtually all production would be channeled into self-maintenance, rather than growth. Odum admitted that some of these hypothetical trends were speculative, but he believed that the overall scheme was universally valid. "While one may well question whether all the trends described are characteristic of all types of ecosystems," he concluded, "there can be little doubt that the net result of community actions is symbiosis, nutrient conservation, stability, a decrease in entropy, and an increase in information."[85]

Strategy also had another meaning in Odum's paper. Nature might not have goals, but humans do. Odum complained that the social and economic strategies pursued in industrialized societies were shortsighted and in conflict with natural ecological processes. A decade earlier, he and his brother had contrasted the natural stability of the coral reef at Eniwetok with the chaos produced by World War II.[86] By the end of the 1960s natural models for human society seemed even more appropriate. The successional trends that he had identified with stable, mature ecosystems also served as guides for rational agricultural and industrial development. The strategy of ecosystem development was also a prudent strategy for human social development.

Comparing nature's strategies with human strategies was a clever literary device. Even without its set of concrete predictions about the effects of succession, Odum's paper would have been memorable as a statement of environmental principles. But by the end of the 1960s many ecologists were becoming impatient with talk of evolutionary or ecological strategies. It sounded perversely teleological, and it could be seriously misleading. The initial volleys of this dispute had actually occurred several years earlier at the 1959 meeting of the Ecological Society of America. During a symposium on energy flow, Howard Odum introduced his electrical circuit diagrams of ecosystems. In the ecological case, Odum claimed that energy was driven by an "ecoforce" analogous to voltage in the electrical circuit. This, according to Odum, necessitated a fundamental change in the way ecologists thought about predator-prey relationships. "The validity of this application [Ohm's Law] may be recognized," he asserted, "when one breaks away from the habit of thinking that a fish or a bear catches food and thinks instead that accumulated food by its concentration practically forces food through the consumers."[87] This rather nonbiological interpretation of predation apparently struck many in the audience as ludicrous. Odum, himself, later admitted that it had alienated him from many ecologists.[88]

This early episode highlighted a serious intellectual problem in ecosystem ecology. Treating trophic levels as black boxes with energy inputs and outputs had proved to be a powerful investigative tool, but when pushed too far it could also mislead. Thermodynamically, there might be nothing wrong with the idea of prey being forced down the throats of predators, but biologically there certainly was. The living world simply did not operate that way. Although less extreme, Eugene Odum's "Strategy of Ecosystem Development" suffered from the same defect. The idea of strategy was obviously a metaphor, but it seemed to suggest that ecosystems had goals and that the parts of the system played functional roles toward achieving those goals. Not only did this imply teleology, but it also ignored fundamental evolutionary principles.[89] By the end of the 1960s most evolutionary biologists believed that natural selection operated upon individuals, not upon collections of individuals. Even metaphorically, ecosystems could not have strategies.

The "Strategy of Ecosystem Development" was frequently cited, but it was also widely criticized.[90] It appeared during an important time of transition in ecology. By the end of the 1960s ecosystem ecology was a well-established specialty, but it was facing increasing opposition from other groups of biologists. Contrary to Eugene Odum's hope that all ecologists would rally around the ecosystem, the discipline was becoming increasingly divided. The sources of this conflict are explored more fully in the next two chapters.

8

Evolutionary Heresies

Evolutionary thinking concentrates attention on the behaviour of the individual and his descendents. If nothing in biology has meaning except in the light of evolution and if evolution is about individuals and their descendents—i.e. fitness—we should not expect to reach any depth of understanding from studies that are based at the level of the super-individual.

—JOHN L. HARPER, "The Contributions of Terrestrial Plant Studies to the Development of the Theory of Ecology"

Perhaps today's theory of natural selection . . . may not, in any absolute or permanent sense, represent the truth, but I am convinced that it is the light and the way.

—GEORGE C. WILLIAMS, *Adaptation and Natural Selection*

BY THE MID-1960s ecosystem ecology was a well-established specialty. The idea of the ecosystem had developed into a broad concept applicable to a wide range of natural phenomena. Ecosystem research revolved around a set of problems that was widely recognized as being both intellectually exciting and practically significant. The leading college textbook of ecology was strongly oriented toward the study of ecosystems. Several successful research programs, established at various institutions around the United States, were attracting substantial financial support from outside funding agencies, both public and private. By any standards, the specialty was a success. But during the early period of growth, ecosystem ecologists aimed at more ambitious goals. Ecosystem ecology was to be more than simply one of several ecological specialties; it was to be the nucleus of a unified ecological science.[1] This was what Eugene Odum meant when he spoke of the "new ecology."

Odum may not have been speaking for a majority of ecologists; not all ecologists were so enthusiastic about ecosystem studies, but the early 1960s was a period of tremendous optimism within ecology. For many, the possibility of a unified theoretical ecology seemed within grasp.[2] The exact nature of such a general ecological theory was not always clear; however, a diverse group of prominent ecologists agreed rather broadly that phenomena at the levels of the population, community, and ecosystem could find a common explanation. This theoretical synthesis would not only unify a rather heterogeneous discipline, but it would also allow ecologists to solve pressing environmental problems. By the end of the decade this optimism had evaporated. Theoretical ecology had become strongly polarized by two groups: ecosystem ecologists and a self-consciously assertive group of evolutionary ecologists whose primary interest was studying populations. Why should the 1960s have been such an important time of transition in theoretical ecology? In this chapter I describe the shifting intellectual landscape of this fractious decade. And in chapter 9, I discuss some institutional and broader social factors that contributed to this intellectual transition.

Sources of Intellectual Conflict

Ironically, the controversies of the late 1960s partly reflect ecology's very success as a discipline. During the decades following World War II, the number of ecologists, both in the United States and abroad, increased dramatically. When Raymond Lindeman's trophic-dynamic paper was published in 1942 the Ecological Society of America had barely 700 members. By the time Odum proclaimed the arrival of the new ecology in 1964, the society's membership was nearly 3,000.[3] Such growth allowed, indeed compelled, ecologists to become specialists. As the number of professional ecologists increased, small, informal networks of researchers sharing common interests began to form. At the same time, the rapid expansion of the scientific literature in the field forced ecologists to narrow the scope of their professional interests. This change was often quite dramatic, and it quickly set the stage for rancorous disputes between different groups of specialists.[4]

More than growth and specialization was involved, however. By the mid-1960s a number of ecologists found ecosystem ecology a bit stodgy, if not misguided. Energy flow, biogeochemical cycling, and succession, though far from completely understood, were well-established areas of research. For an ambitious young ecologist looking to stake an intellectual claim other research areas appeared more attractive. There was a growing belief among many younger ecologists that evolutionary theory had been ignored or misunderstood by their elders. Earlier ecologists, it was widely believed, had used evolutionary ideas in vague and uncritical ways. Ecosystem ecologists were not the only culprits, but they were prominent offenders,[5] often suggesting that natural selection acted at all levels of biological organization. They commonly implied that evolution was a directional process and tended to conflate the processes of evolution and succession. Some ecosystem ecologists seemed to believe that evolution was not so much the ultimate cause of patterns and processes in nature, but an epiphenomenon of energy flow. Not surprisingly, when a few critics began to call for a more "Darwinian approach" to ecology they found a receptive audience among younger ecologists imbued with the modern evolutionary synthesis.[6]

The theoretical synthesis of Darwinism and Mendelism, which began in the 1920s, was largely accomplished by the end of World War II.[7] The postwar period saw the expansion and consolidation of this new body of evolutionary theory. Later, it invaded a number of fields that had contributed little to the original synthesis, notably ecology. The new evolutionary ecologists who championed the modern synthesis brought a critical attitude to bear on ecological theory, and they attempted to eliminate rival interpretations.[8] Some critics, suggesting that certain evolutionary ideas were quickly accepted, with little dissent, by the biological community, have portrayed this process of theoretical consolidation as a "hardening of the synthesis."[9] Whether or not one accepts this as a broad historical generalization, it is certainly true that by 1970 being an evolutionary ecologist meant something quite different than it had ten years earlier.

The most fundamental change in evolutionary ecology came in the way that ecologists used the concept of natural selection. Prior to about 1965, it was widely believed that natural selection acted at a number of levels of biological organization.[10] Commonly, scientists claimed that selection acted for "the good of the species" or perhaps even for the entire ecosystem. Such group selectionist explanations were often implicit or simply uncritically stated, but in 1962 the British ecologist V. C. Wynne-Edwards published a detailed defense of the concept.[11] The controversy that Wynne-Edwards's book engendered is discussed in detail later in this chapter. Suffice it to say, however, that the debate ended with "quick and brutal finality."[12] By the time that Richard Lewontin wrote a thoughtful article on the levels of selection in 1970, it was generally accepted by evolutionary ecologists that natural selection operated primarily at the level of the individual.[13] Group selection, if it occurred at all, did so only under highly restricted conditions.

Accompanying this more narrowly focused interpretation of natural selection was the rise of what might be called "evolutionary reductionism." Because selection always, or almost always, operates at the level of the individual, evolutionary ecology was drawn increasingly toward the study of populations. By 1965 a rich body of theoretical and empirical studies on populations could provide the foundation for a new approach to evolutionary ecology.[14] The situation was ripe for aggressive rhetoric and programmatic statements. Speaking to an audience of some three hundred biologists at Syracuse University in the spring of 1967, Richard Lewontin decried the discredited holism of earlier ecological and evolutionary research.[15] Practitioners of the old approach were the "stamp collectors" of biology, fuzzy-headed thinkers incapable of analyzing complex biological problems. According to Lewontin, the future of evolutionary biology rested on the same reductionist methodology used so successfully in molecular biology: simplifying complex systems into constituent parts, building theoretical models to determine the limits of possibility, and conducting rigorous experiments—often in the laboratory.

The tone of Lewontin's comments was inflammatory, and in later years he became much more critical of reductionism. But at the time, his aggressive optimism carried conviction. A year earlier, he had published an influential pair of articles showing how the tools of molecular biology could be applied to evolutionary problems.[16] Using electrophoresis, a method for separating polypeptides by their electric charges, he estimated the genetic diversity in natural populations of fruit flies. The implications of Lewontin's research were not lost on the ecologists in the audience. Genetics held the key to understanding evolution, and evolutionary ecologists needed to become more knowledgeable about the genetic basis of the traits they studied.[17] Perhaps, a new interdisciplinary field, population biology, might arise out of the fusion of evolutionary ecology and evolutionary genetics. If so, this new area of research would probably have little in common with ecosystem ecology. To the extent that there was a connection, however, population biology served as the intellectual foundation for ecosystem ecology, not vice versa. According to two leading advocates of population biology—a geneticist and an ecologist—"we strongly believe that before a satisfactory understanding of ecosystem structure can be achieved, it is helpful, if not mandatory, to acquire a better knowledge of the structure of populations and their interactions with the environment."[18]

The growing ties between genetics and ecology may have contributed to the rift between ecosystem ecology and the new evolutionary ecology, but reductionism drove the wedge between the two ecological specialties. In his early discussions of holism and reductionism, Eugene Odum rarely emphasized the philosophical differences among ecologists. The reductionism of cellular and molecular biology most concerned him. Later, he increasingly turned his arguments toward the type of evolutionary reductionism expressed by Lewontin in his Syracuse address.[19] The ensuing debates over holism and reductionism in ecology were often rancorous and not very enlightening.[20] They did, however, highlight a growing intellectual rift within the discipline.

The distinction between evolutionary ecology and ecosystem ecology was further exaggerated by an influential philosophical article that appeared in *Science* magazine in 1961.[21] In "Cause and Effect in Biology" Ernst Mayr distinguished between *proximate causes,* the immediate physiological or behavioral causes of a biological phenomenon, and *ultimate causes,* the historical or evolutionary causes of the phenomenon. He also distinguished between a *functional biology* and an *evolutionary biology,* based upon the two respective forms of causation. The proximate-ultimate dichotomy had been used earlier, but Mayr's clear exposition and the wide readership of *Science* insured that the distinction would become a commonplace of philosophy of biology. Ecologists quickly adopted Mayr's terminology.[22]

Despite the clarity of Mayr's writing, there were important ambiguities in his argument. Were functional and evolutionary biologies simply alternative approaches to research, or were they distinct biological fields? Were they complementary, or did they embody two very different philosophies of biology? Could an individual scientist be both an "evolutionary biologist" and a "functional biologist," and, if so, could one be both at the same time? Some ecologists stressed the complementarity of functional and evolutionary biologies, but more often the distinction emphasized the incommensurability of two schools of ecological thought.[23] Ecosystem ecology, a form of functional biology, became by implication nonevolutionary.

This was more than a rhetorical ploy; a real issue was at stake. "Function" was at the very heart of the ecosystem concept, and critics were justified in pointing out the evolutionary problems with functional explanations. Running through the ecological writings of

Forbes, Elton, Lindeman, and the Odum brothers was the notion that species play functional roles within the community or ecosystem. They are parts of the system, much as gears are parts of a clock or organs are parts of an organism. The organismal and mechanical metaphors used so frequently by these ecologists both reflected and reinforced this way of thinking about nature.[24] Take, for example, the role of bacteria and other microorganisms in decomposition. Without decomposition nutrients would become completely tied up in dead organic material, biogeochemical cycling would cease, and life on earth would slowly grind to a halt. Thus, the integrity of the ecosystem depends upon the activities of these tiny organisms. It is easy enough to think of decomposition as the function of the microorganisms; that is, they are the decomposers in the system. The problem arises when one slips into the habit of thinking that decomposers evolved for this purpose. There was a tendency for ecosystem ecologists to at least imply that this was the case.[25] As critics pointed out, however, nutrient recycling is simply a fortuitous consequence of the physiology of these microorganisms, a physiology that has evolved to benefit the bacteria, not to improve the efficiency of the ecosystem. Digesting dead organic material to obtain energy is a function of bacteria, but recycling inorganic nutrients to plants is probably not. Ecosystem ecologists were slow to recognize this subtle distinction, and they became easy targets during the group selection debate. This inertia reinforced the image of ecosystem ecology as a nonevolutionary science.[26]

The Group Selection Controversy

Using hindsight one can classify earlier generations of evolutionary biologists as "group selectionists" or "individual selectionists," but this exercise is not particularly enlightening. Prior to the early 1960s many evolutionary biologists were only dimly aware of the concept of units of selection that so preoccupied later evolutionists.[27] Beginning about 1960, however, a small group of evolutionary biologists began carefully to distinguish various levels at which selection could operate, and they explicitly argued that selection did, in fact, operate at levels above that of the individual. Timing can be critical in science. Although these ideas were not new, group selectionist arguments met with a storm of criticism when they were presented during the early 1960s.

The most prominent of the group selectionists was the British zoologist V. C. Wynne-Edwards. His *Animal Dispersion in Relation to Social Behaviour* (1962) ignited the controversy over levels of selection, and it became the major target of criticism against group selection. Wynne-Edwards was educated at Oxford University. For sixteen years he taught at McGill University (1930–1946), before returning to Britain as professor of zoology at Aberdeen University, a position he held for the rest of his career.

Wynne-Edwards wrote his book in response to David Lack's *The Natural Regulation of Animal Numbers* (1954). He accepted Lack's argument that population size is maintained by density-dependent factors. In other words, the influence of factors regulating population growth varies with the degree of crowding in the population. He also agreed with Lack that these regulatory mechanisms are the product of natural selection. Where the two ornithologists disagreed was the level at which selection operates. Wynne-Edwards believed that density-dependent regulation was a group adaptation, a kind of homeostatic system that had evolved through natural selection at the level of the population. In other words, over the course of evolutionary time more homeostatic populations tended to replace less homeostatic ones. Lack also considered population regulation to be analogous to homeostasis, but as his critique of Wynne-Edwards's book made clear, he believed that population regulation could be explained completely in terms of individual selection.[28] For Lack to say that a population was homeostatic meant nothing more than saying that in a statistical sense individuals in the population tended to have the optimum number of offspring. It was not the population that was selected, but the individuals within the population.

In the introduction to his book, Wynne-Edwards presented his argument for group selection using the analogy of commercial fisheries.[29] Theoretically, a commercially valuable species can be exploited at some optimal level; below this level profits decrease, but above this level the fish population is depleted, endangering future harvests. Although fishery biologists can estimate this optimal level of exploitation, politically it is difficult enforce fishing limitations. Self-interest and unchecked competition must be restricted for the common good, but there is always an economic incentive for individuals to cheat. According to Wynne-Edwards this case illustrated a number of general regulatory principles applicable to nature as well as to human activities. No profitable, renewable resource is immune from overexploitation. All participants must agree to limit exploitation, and such universal limits must be agreed upon before the resource begins to decline. Finally, individual agreement is not a sufficient guarantee of compliance; some cheating always occurs. Therefore, enforcement of the regulatory program requires the existence of some "higher court."[30]

In nature, this higher court was group selection. Every population has the potential for exponential growth, and because growth cannot be controlled instantaneously, every population is capable of depleting its resources. But this rarely occurs; cases of mass starvation in nature are quite rare. Evidently, population growth is checked even before resources begin to decline. According to Wynne-Edwards, this exquisite form of self-regulation has evolved through natural selection acting upon populations. Populations whose self-regulatory machinery is inadequate go extinct. Self-regulating populations persist and replace their less successful competitors. Through this process of selection the mechanisms of self-regulation are gradually perfected. In the words of ecologist Lawrence Slobodkin, evolution tends to produce populations of "prudent predators."[31]

Wynne-Edwards did not simply describe group selection; he also provided a detailed discussion of the mechanisms by which it might actually work in various populations. Food was the ultimate factor limiting population size, but open competition for food, which might lead to bloodshed and death, had generally been replaced by what Wynne-Edwards referred to as *epideictic displays.*[32] Singing in birds, claw displays in fiddler crabs, and the expanded fins of Siamese fighting fish were examples of such displays. These symbolic behaviors acted as regulatory signs and signals; information about population density and resource abundance were transmitted from one individual to another. As a result, physiological and behavioral changes occurred within breeding individuals to increase or decrease fecundity. If the population were too large, then individuals produced fewer offspring; if the population were too small, then individuals produced more. Thus, epideictic displays served as the proximate causes of population regulation.

This was a functionalist argument. Individuals regulated their reproductive output to optimize the overall size of the population. Thus, sometimes individuals sacrificed their own evolutionary fitness to benefit the population as a whole. This self-sacrifice was regulated and enforced by the higher court of group selection. Natural selection at the level of the individual might favor reproductive cheating, but group selection always eliminated populations in which selfishness flourished: "Where the two [forms of natural selection] conflict, as they do when the short-term advantage of the individual undermines the future safety of the race, group-selection is bound to win, because the race will suffer and decline, and be supplanted by another in which antisocial advancement of the individual is more rigidly inhibited."[33]

The possibility that "social parasitism" might evolve through individual selection had long troubled evolutionary biologists, including some architects of the modern synthesis.[34] Group selection provided an attractive solution to the problem, particularly for biologists concerned about the implications of evolutionary theory for human society. For example, the University of Chicago ecologists Warder Clyde Allee and Alfred E. Emerson used group selection as not only a biological theory but also a basis for human ethics.[35] Wynne-Edwards did not discuss social theory at great length, but he shared the concerns of Allee and Emerson. In the absence of regulation, ruinous exploitation in the fishing industry was inevitable. "What is fundamental to conservation is that the system of free enterprise, with every man or company for himself, must be exchanged for a common code of rules," he wrote.[36] Rational cooperation in human societies held out the possibility of regulating selfish behavior; humans could truly be prudent predators. This was impossible in other species, and Wynne-Edwards criticized Allee for believing in a nebulous principle of animal cooperation.[37] Cooperation among animals was merely the automatic result of the differential survival and reproduction of populations. Selfish populations disappeared, being replaced by their more cooperative neighbors.

Although the mechanism was different, the results of human rationality and group selection were much the same. The higher court of group selection effectively restrained free enterprise in the animal world just as rational governance did in human societies: "conventions [i.e., epideictic displays] lay down codes of law, which have evolved to safeguard the general welfare and survival of the society, especially against the antisocial, subversive self-advancement of the individual."[38] Animals were neither ethical nor prudent, but group selection led to something analogous to these human characteristics. In successful populations predators did not exterminate their prey. Through the complex workings of the behavioral feedback system the population regulated its size so that the food supply would not be exhausted.

The anthropomorphism that characterized part of his discussion of group selection strongly suggests that Wynne-Edwards was concerned about the behavior of the human animal as well as that of bird and fish. But analogies with human society were only a minor element in Wynne-Edwards's argument. Underlying the entire discussion were the twin concepts of homeostasis and negative feedback. For Wynne-Edwards, like so many other ecologists of his generation, populations were organic entities. The system of epideictic displays provided a cybernetic mechanism for controlling population growth. When the population was small and resources abundant, these behavioral signals stimulated reproduction. As the population approached its optimum size, behavioral changes curtailed reproduction. Population regulation was dynamic; the size of populations constantly fluctuated in response to the availability of environmental resources, but social behavior tended to optimize population size. This regulatory system was mechanical; the "instrumentation" of social behavior monitored population growth much the way a thermostat regulates temperature.[39] To an even greater extent, however, population regulation was physiological. "To build up and preserve a favorable balance between population-density and available resources," Wynne-Edwards claimed, "it would be necessary for the animals to evolve a control system in many respects analogous to the physiological systems that regulate the internal environment of the body and adjust it to meet changing needs. Such systems are said to be homeostatic or self-balancing, and it will be convenient for us to use the same word."[40]

Homeostasis had become a ubiquitous concept in ecology and evolutionary biology, but Wynne-Edwards's book, with its controversial claims about group selection, focused critical attention on the idea. Critics denied that population regulation was analogous to homeostatic control, and some ridiculed the idea that populations, communities, or ecosystems shared any important characteristics with machines or organisms. Charles Elton, who as a young man had often employed such analogies, now complained that Wynne-Edwards's use of group homeostasis was a "sweeping arm-chair notion."[41] David Lack admitted that population regulation might be analogous to homeostasis in organisms, but in fact his views were diametrically opposed to those of Wynne-Edwards.[42] For Wynne-Edwards, populations were organic entities; individuals were functional parts of the larger whole. As such, individuals might sacrifice individual fitness to benefit the group, and this altruism was enforced by natural selection acting on populations as units. For Lack, populations were simply aggregations of individuals, with each individual acting in its own self-interest. Populations were regulated, but not by the kind of complex, negative feedback system suggested by Wynne-Edwards. Regulation came about completely as a result of differential survival and reproduction of individuals.

Lack was only one of a number of biologists who criticized Wynne-Edwards's theory in short reviews. However, the most devastating attack on group selection was George C. Williams's *Adaptation and Natural Selection* (1966). Unlike Lack and Elton, who were about the same age as Wynne-Edwards, Williams became a biologist after World War II. He was a member of the first generation of students trained in the era of the modern synthesis. R. A. Fisher, J.B.S. Haldane, and Sewell Wright were his intellectual models. These men had pointed the way toward a new evolutionary biology, but their ideas were still sometimes ignored or misused. Therefore, Williams saw his book as more than a critique of Wynne-Edwards's ideas; it was also an aggressive defense of the new approach to evolutionary ecology. According to Williams, he intended to "purge" biology of unnecessary impediments to Darwinian theory.[43]

It would be difficult to find two more dissimilar books than those of Wynne-Edwards and Williams. Wynne-Edwards's *Animal Dispersion* was a ponderous work, more than 650 pages, filled with examples supposedly illustrating the behavioral mechanisms of group selection. His general ideas were presented quite clearly in an introductory chapter, but it seems unlikely that many biologists read the book from cover to cover. Williams's *Adaptation and Natural Selection* was only half as long. There were a few well-chosen examples from nature, but the book was primarily a logical defense of individual selection. It was a tremendous success; within a decade the book went through five hardcover printings and a paperback edition.

Williams's book launched a three-part attack on what he considered to be serious errors in evolutionary thought. First, he proposed more rigorous and restrictive definitions for two related terms: function and adaptation. An *adaptation* was a means or mechanism fashioned by natural selection for a particular biological purpose or goal—that is, its *function*.[44] Thus, according to Williams, the use of these terms always implied evolutionary cause and effect. In other words, Williams was proposing to shift the meaning of function from the traditional definition based primarily upon proximate causation, to an evolutionary definition based primarily upon ultimate causation. Under this new definition many functional explanations became invalid, and much functional biology, particularly functional ecology, became suspect. For Williams, incidental or fortuitous effects, however beneficial, could not be functions. The function of glycolytic enzymes in a yeast cell is energy transformation. Production of alcohol, which may benefit humans, is an incidental effect of this transformation process.

Therefore, one cannot properly speak of alcohol production as a function of yeast physiology. Similarly, one could not ascribe functional roles to members of a population or ecosystem, unless one could clearly demonstrate that the activities had evolved specifically for that purpose. From Williams's point of view, the recycling of nutrients to the plants in an ecosystem is not a function of decomposing bacteria; it is simply a fortuitous side effect of bacterial metabolism.

The second part of Williams's critique centered on the related concept of *group adaptation*. Group selectionists had argued that certain adaptations were the properties of groups rather than individuals. But in most cases, Williams argued, supposed group adaptations could be adequately explained as the summation of individual adaptations or as effects incidental to these individual adaptations.[45] For example, during cold weather groups of small mammals often huddle together, each individual contributing heat to the group. Superficially, this appears to be a group adaptation for maintaining collective warmth, but, as Williams pointed out, by huddling each individual is insulating itself. Therefore, this social behavior is not a group adaptation, but rather an incidental result of individual behaviors. From family groups to ecosystems, Williams argued, most purported cases of group adaptation could be similarly explained away. The idea of group adaptation, which permeated many earlier evolutionary discussions in ecology, now appeared superfluous.

At the heart of the group selection controversy was the nature of altruism. Wynne-Edwards and other group selectionists believed that not only did individuals contribute to the welfare of the group but that they also sometimes did so in self-sacrificing ways. Specifically, Wynne-Edwards had claimed that individuals would sacrifice their own reproductive success to optimize overall population size. Relying heavily upon the earlier work of the British theoretician W. D. Hamilton, Williams argued that this was probably never the case. Altruism was not dead, but as E. O. Wilson later remarked, Hamilton had taken most of the good will out of the concept.[46] All behavior, even cooperation and altruism, was basically selfish. For example, group selectionists had sometimes pointed to mammary glands as an example of structures that had evolved to benefit offspring rather than the females possessing them.[47] But as Williams pointed out, mammary glands clearly contributed to the fitness of individual females.[48] By feeding her young, a female ensured that her genes would be represented in the next generation. Other cases of altruism could also be explained in this way. Fitness, in Hamilton's calculus, involved genes contributed to the next generation both directly by one's own reproductive efforts and indirectly by one's close relatives. As Wynne-Edwards himself had admitted, many social groups were made up of close relatives. By sacrificing for other members of a closely related group, an individual was in fact adding to his or her own fitness.

Williams did not bother to refute all of Wynne-Edwards's examples; he did not have to. He had questioned the very concept of group adaptation and shifted the burden of proof to those who claimed that social behavior functioned for anything other than individual self-interest. Finally, he had invoked a "principle of parsimony" in defense of individual selection. In cases where biological effects could be explained in purely physical or chemical terms, the concept of adaptation should not be invoked. Likewise, in cases where adaptation could be unequivocally demonstrated, it ought to be explained in terms of individual selection unless this type of explanation was clearly inadequate. Only then should group selection be invoked. Williams left his readers with the impression that such situations were rare, perhaps nonexistent.

In 1970 Richard Lewontin wrote a thorough review article on the units of selection.[49] In this article he discussed the theoretical conditions for natural selection acting at levels ranging from genes to ecosystems. Selection acting above the level of the individual was not theoretically impossible, but the conditions for this to occur were so unusual that it seemed unlikely that group selection constituted an important evolutionary mechanism. Furthermore, Lewontin showed that group selection could not, in most cases, counteract individual selection. Contrary to Wynne-Edwards's claims, when individual selection conflicted with group selection, the former almost always prevailed. Lewontin's article marked the end of the controversy. Group selection had been widely, if not quite universally, rejected. To be a group selectionist in 1970 removed one from the mainstream of evolutionary biology. This had important repercussions for the relationship between ecosystem ecology and evolutionary ecology.

The Evolution of Ecosystems

The group selection debate was primarily concerned with populations. However, the same type of argument employed by Wynne-Edwards was also frequently used to explain the evolution of ecosystems. Typical of this theoretical approach was the work of Maxwell Dunbar, a marine biologist and student of Wynne-Edwards. Born in

Edinburgh, Scotland, in 1914, Dunbar earned bachelor's and master's degrees from Oxford University. There he was strongly influenced by Charles Elton's ideas on population cycles and community ecology.[50] As a student, he spent a year (1937–1938) on a Henry Fellowship at Yale. During this year he worked closely with G. Evelyn Hutchinson and his students. Hutchinson's approach to research made a deep impression on the young biologist, and it was the Yale ecologist who later encouraged Dunbar to publish his theory of ecosystem selection in the *American Naturalist*.[51] In 1941 Dunbar received his Ph.D. from McGill University under Wynne-Edwards. The two men remained on friendly terms, although Dunbar later claimed that his teacher had little influence upon his evolutionary ideas. During the war years, Dunbar served as Canadian consul to Greenland. He had traveled to the arctic a number of times as a student, and the consular position provided an opportunity to continue his studies in arctic marine biology. After the war he returned to teach in the zoology department at McGill University.

His research on the biology of arctic oceans impressed upon Dunbar the great differences between tropical and polar ecosystems. In contrast to the stability of ecosystems near the equator, those in the arctic were often characterized by destabilizing oscillations. According to Dunbar, the constancy of tropical environments allowed ecosystems to evolve toward a stable, steady state; indeed, tropical ecosystems had nearly reached the limit of stability. This process occurred much more slowly and imperfectly in the fluctuating environment of the arctic. "An absolutely stable system, if one can imagine such a thing in the living world, can no longer change or adapt," Dunbar claimed. "But the ecosystems of the mid and higher latitudes have not yet come anywhere near to such a condition. Hence one must expect the evolution of more species, more niches, greater specialization, up to the maximum possible as . . . has been reached in the wet tropics."[52] This evolutionary process occurred through natural selection acting upon entire ecosystems. "As to the mechanisms by which selection might take effect at this level," Dunbar wrote, "they are of the ordinary Darwinian sort except that the criterion for selection is survival of the system rather than of the individual or even the species."[53]

Dunbar premised his argument on the widely accepted idea that oscillations in any biological system are maladaptive. In other words, oscillations are the antithesis of stability. Destabilizing oscillations may have far-reaching biological effects. For example, population cycles carry the possibility of extinction for the population, but they may also destabilize the living community leading to local extinction of the entire ecosystem. When such extinctions occur, empty environmental spaces are created that can then be colonized by adjacent ecosystems. The more stable the ecosystem, the more likely it will persist, and the more successful it will be in colonizing new areas.

Similar ideas had been expressed by other ecosystem ecologists, most notably, Ramon Margalef and the Odum brothers. Dunbar's argument, however, was more explicitly Darwinian than the others; it contained all the elements of natural selection. There was variation in the stability of communities and ecosystems. This variation was heritable in the sense that stable communities and ecosystems replicated themselves by colonizing adjacent areas. And selection occurred among these replicates: stable ecosystems persist and colonize new areas; less stable systems go extinct. Like other supporters of group selection, Dunbar believed that selection at higher levels of organization could override individual selection. For example, one could imagine a situation where high fecundity might be advantageous to the individual but destabilizing for the ecosystem as a whole. In such situations, selection among ecosystems might eventually lead to lower fecundity. If individuals continued to reproduce beyond the optimum number, then the population might destroy its habitat, the entire ecosystem might collapse, and the living community might then be replaced by colonizers from a better adapted neighbor. In this way selection among ecosystems might lead to lower reproductive rates, even in the face of individual selection for higher fecundity.

Dunbar deplored the narrow-minded emphasis on individual selection in contemporary evolutionary theory.[54] Nonetheless, he considered himself a Darwinian. It is important to emphasize this point because, although the group selection controversy was about mechanisms of evolution, it was not a controversy between "Darwinians" and "anti-Darwinians." Both group selectionists and individual selectionists believed in natural selection. Only the level at which the process occurs was at issue. Perhaps that is why the debate became so bitter; heretics, rather than nonbelievers, are the greater threat to orthodoxy.[55]

Dunbar's theory of ecosystem evolution was only briefly discussed and criticized in Williams's *Adaptation and Natural Selection.*[56] For Williams, it was part and parcel of Wynne-Edwards's discredited thesis; the idea that individuals might curtail reproduction in order to stabilize the ecosystem was gratuitous and unnecessary. Many ecologists agreed, and Williams's critique had serious repercussions for the "new ecology" that Eugene Odum and Howard Odum had been building so assiduously. Like Wynne-Edwards and Dunbar, the Odums consistently used functional arguments. Individuals really were parts of populations, and populations were parts of ecosystems. These parts were similar to the components of the electronic circuits that Howard Odum used as models for ecosystems. Like resistors or capacitors, they played functional roles in the system, one of which was homeostasis or maintaining the stability of the whole. "Survival of the fittest" with its implication of a "dog-eat-dog" world, an idea Eugene Odum thought was at the heart of individual selection, was anathema to his belief in highly structured ecosystems characterized by pervasive interdependence.[57] In stark contrast to the evolutionary reductionism of Williams's book, where the emphasis was upon individual adaptation and selection, the focus of attention in the Odums's ecology was almost always on the entire ecosystem and its persistence as a unified whole. For Williams it would be inconceivable to speak of ecosystems having evolutionary strategies—even as a metaphor. For Eugene Odum it was difficult not to think in those terms.[58]

One cannot accuse Odum of vacillation or inconsistency. While most ecologists were turning against group selection, he rallied to its defense. In his early writings, Odum had used the concept of natural selection rather uncritically. Like most ecologists he had not carefully distinguished among the levels of selection. After the group selection controversy he did, and by the early 1970s, he was using group selection as part of his explanation for ecosystem stability.[59] According to Odum, stability evolves in two ways. Symbiosis, Darwin's web of complex relations, could evolve through coevolution. Odum had long argued that what began as parasitic relationships often evolved into mutually beneficial ones. As mutualistic partners cooperate and become increasingly dependent upon one another, the ecosystem of which they are a part is stabilized. This part of Odum's argument did not necessarily require group selection. Coevolution can be explained completely in terms of individual selection; two individuals may cooperate, but each does so to increase its own individual fitness.

Coevolution was important, but by itself it could not account for the stability of ecosystems. According to Odum, cooperation almost always involves self-sacrifice, at least initially. Selfishness may have immediate rewards, but in the long run cooperative groups are more successful than selfish ones. Through a kind of evolutionary trial and error, group selection eventually eliminates the selfish groups. The evolution of cooperation is more likely to occur if populations are subdivided into small groups of closely related individuals, but Odum believed that once the process began, populations would become so interdependent that cooperation would become a general characteristic of the living community as a whole. Natural selection would then begin to discriminate among populations and communities. Those with greater stability—that is, more cooperation—would tend to persist by replacing their less stable neighbors.

Once the dust settled from the controversy over Wynne-Edwards's book, a few evolutionary theorists began to take a second look at group selection, albeit in a much more restricted sense than that used earlier by the British ornithologist. For some, a rigorous theory of group selection might provide an intellectual bridge between the functional ecology of Eugene Odum and contemporary evolutionary theory.[60] Odum later used this literature to justify his position, but to little avail. Critics continued to attack the functionalist assumptions and the evolutionary misconceptions that seemed to be at the heart of the ecosystem concept.[61] Odum continued to write about ecosystem evolution, but many other ecosystem ecologists simply retreated from the fray. Evolution obviously had something to do with ecosystems, but one could do interesting research on ecosystems without ever mentioning evolution.

The Death of Optimism

The group selection controversy ended in an important, but pyrrhic, victory for evolutionary ecologists. The rise of evolutionary ecology came at the expense of the intellectual polarization of theoretical ecology. In 1960 ecologists were optimistic that a general theory of ecology was imminent. Such a theory would apply to populations, communities, and ecosystems; it would explain both evolution and energy flow; and it would solve problems both pure and applied.[62] The theoretical controversies of the decade eroded this optimism. By 1970 few ecologists were talking about grand unifying theories. Ecosystem ecologists complained that evolutionary ecologists ignored levels of organization above the population. Evolutionary ecologists complained that ecosystem ecologists ignored or misunderstood commonly accepted evolutionary ideas. A great gulf seemed to divide the two approaches. Those evolutionary ecologists who ventured into ecosystem studies found an intellectual wilderness. Ecosystems might have something to do with evolution, but energy flow, nutrient cycling, succession, and other ecosystem level phenomena seemed so far removed from evolutionary theory that most evolutionary ecologists found it difficult even to visualize their role in studying them.[63] The effort seemed hardly worthwhile. Ecosystem ecology and evolutionary ecology had parted company.

The erosion in optimism that accompanied this split had far-reaching effects on ecology. It came just at the time that ecologists were being called upon to provide answers to a growing list of environmental problems. Ecosystem ecologists, particularly the Odums, had emphasized the important role that ecologists could play in shaping public policy. This had been an important theme running through all the editions of Eugene Odum's popular *Fundamentals of Ecology.* The textbooks that began to replace Odum's during the 1970s exhibited little confidence that ecologists could play an important leadership role in the environmental movement. Many either barely mentioned environmental problems or discussed them only briefly in a chapter or two. Typical was Robert Ricklefs's highly successful *Ecology* (1973), the most encyclopedic textbook of the decade.[64] Only the introductory chapter discussed the relationship between theoretical ecology and environmental problems, and Ricklefs held out the hope that eventually professional ecology might contribute to solving these problems. But in marked contrast to Odum, whose textbook was filled with examples of how contemporary ecological theory could be applied immediately to environmental problems, Ricklefs rarely discussed applied ecology in the body of his textbook. If Eugene Odum shared his brother's "technocratic optimism," an ambitious belief in the possibility of ecological engineering, the leading textbook writers of the 1970s were neither technocrats nor optimists. Their ambivalent attitude may have reflected the skeptical temper of the 1970s, but it also reflected the fact that by this time professional ecologists were deeply divided. Ecosystem ecology provided the ideal perspective for examining critical environmental problems, but for many evolutionary ecologists this perspective lacked an acceptable intellectual foundation.

9

Big Ecology

Big Science is an inevitable stage in the development of science and, for better or for worse, it is here to stay.

—ALVIN M. WEINBERG, "Impact of Large-Scale Science on the United States"

Ecologists will have to learn to curb some of their traditional individuality, to learn how to work in large teams harmoniously and effectively. . . . What is needed is nothing less than a new psychology or a new sociology for ecologists.

—STANLEY AUERBACH, congressional testimony, 1968

And, tucked away in some academic corners, modern Big Science probably contains shoestring operations by unknown pioneers who are starting lines of research that will be of decisive interest by 1975.

—DEREK J. DE SOLLA PRICE, *Little Science, Big Science*

BY THE LATE 1960s ecology was beset by a deep intellectual schism. This division were exacerbated by an extraordinary event that changed, at least temporarily, the way that ecological research was organized in the United States. From 1968 to 1974 the United States actively participated in the International Biological Program (IBP). Inspired by the theme, "The Biological Basis of Productivity and Human Welfare," the IBP had a strong orientation toward ecological research, specifically ecosystem studies. Some 1,800 American scientists participated in various research projects. These large projects were supported primarily by the National Science Foundation (NSF), which, together with other federal agencies, pumped between $40 and $60 million into the ecological programs of IBP.[1] During the peak year of 1973, ecologists in the United States received close to $11 million dollars in support of IBP-related projects. This represented unprecedented support for ecological research. Under different historical circumstances it might have provided a springboard for the unification of the discipline. But this was not to be. Ecologists themselves were not unanimous in their support of "big ecology." Some ecologists, primarily evolutionary ecologists, felt shortchanged by the program. They tended to be just as hostile toward the IBP as the molecular biologists who roundly condemned the new large-scale funding for ecology.

Criticism of the IBP did not diminish after it officially ended in 1974. Opponents, claiming that the program had produced little of value, complained that its worst features had become institutionalized in the new Ecosystem Studies Program at NSF. Although some other ecologists came to the defense of the IBP, few gave it a ringing endorsement. What was the legacy of the IBP? Did it produce any scientific contributions of lasting value, or, to borrow a phrase from conservative social critics of the period, did it "throw money" at ecological problems without solving them? Was it a bold and innovative experiment in coordinating large-scale ecological research, or was it big science in a field better suited to traditional small-scale approaches to research? Before attempting to address these questions, we must examine the philosophy of big science that emerged from World War II.

The Philosophy of Big Science

For many scientists, World War II provided a utopian model for research.[2] The wartime experience gave scientists a great sense of purpose and accomplishment, for they had contributed decisively to victory in the greatest war in history. The Manhattan Project and other large-scale scientific efforts had been the result of teamwork, unprecedented cooperation among theoreticians, experimentalists, and engineers. Finally, the wartime experience seemed to exemplify the virtues of scientific planning; given adequate resources, manpower, and national commitment to a goal, even enormous technical problems could be solved quickly.[3]

After the war most scientists were eager to return to their academic posts and the familiar routines of traditional science. But a number of academic entrepreneurs and scientific planners sought to implement the utopian model of big science in peacetime research. In their enthusiasm, they were not always fully cognizant of the limitations of large-scale scientific research programs. A central premise of this chapter is that big science works well only when a number of unusual social and intellectual conditions apply. The program must be directed toward a well-defined goal, both scientifically interesting and socially compelling. In other words, it must be supported by not only the scientific community but also a broad political consensus. In general, this means that proponents of the project must make a convincing case that national security, prestige, or the welfare of a powerful constituency is at stake. Finally, the costs of doing the research must be so prohibitive that individuals or informal groups of scientists cannot realistically hope to accomplish the research project.

A number of peacetime projects—notably the activities of the International Geophysical Year and the U.S. space program—appear to have met these conditions. But ecosystem ecology during the late 1960s did not. Ecology had become intellectually fragmented, and there was no consensus within the discipline about the importance of studying ecosystems. This was even more true of the scientific community as a whole. Support from leading scientific organizations was lukewarm, and some influential groups actively opposed the project. Thus when it came to selling big ecology to policy makers and the public, ecologists did not form a united front, and their lobbying efforts were often unsuccessful. Despite public concerns and some notable legislative successes, environmental problems never became a highly visible part of the nation's political agenda. Environmentalism lacked the glamour of space travel and the immediacy of cancer. During the late 1960s, politicians, comparing large-scale cancer research to the Manhattan Project, frequently called for a "war on cancer," and they were willing to appropriate hundreds of millions of dollars annually to the fight.[4] Support for ecological research paled in comparison; the IBP may have been "big ecology," but it never made the big leagues in terms of large-scale scientific projects. Among ecologists, unaccustomed to large-scale federal funding, there was little agreement on the need for highly coordinated, cooperative teamwork associated with big science projects. Critics of the IBP complained that the unnecessary big ecology endangered more traditional approaches to ecological research. At the same time, other ecologists argued forcefully for ecosystem research based upon the wartime model of big science.

In his provocative *Environment, Power, and Society* (1971), Howard Odum called for the formation of "Ecosystems Task Forces."[5] These large teams made up of various specialists would study ecosystems in a series of steps. First, taxonomists would collect and identify the constituent species in the system. The goal of this initial phase of research was the publication of systematic keys to be used by ecologists. During the second phase of research, ecologists trained in natural history and autecology would identify the food chains in the ecosystem and group species into trophic levels. Next, systems ecologists would gather data on the metabolism of the ecosystem: the flow of energy and materials through the trophic levels identified during the preceding phase. Once these data were collected, systems models could be created by theoretical ecologists to study the probable outcome of various disturbances or experimental modifications in the system. These models could then be used by engineers to plan, manage, and manipulate the ecosystem to meet various social goals.

Odum paid lip service to scientific autonomy; members of the task force would be able to pursue individual research projects along with their contributions to the team effort. But in actuality, the ecological task force was as rigidly hierarchical as its military model. Each group of investigators was under the supervision of a senior scientist, and the entire operation was directed by an elite group of systems ecologists. Such a scheme might appear idiosyncratic, even megalomaniacal,[6] but it perfectly encapsulated the philosophy of big science that formed the intellectual foundation for the IBP.

Ironically, Howard Odum did not play a major role in the IBP. In the United States, the IBP was oriented around the study of very large ecosystems or biomes: grasslands, deserts, tundra, coniferous forest, deciduous forest, and tropical rain forest. As chairman of the IBP Tropical Biome planning committee, Odum proposed an integrated research program of several tropical rain forests. The project failed to materialize, and tropical biology was conspicuously absent from the final agenda of the IBP in the United States.[7] However, even before the IBP biome studies began, a large-scale tropical study, directed by Odum, was nearing completion. Supported by large contracts with the Atomic Energy Commission and involving nearly one hundred scientists, the project was conducted between 1963 and 1967 in the El Verde forest of Puerto Rico. Philosophically, if not administratively, Odum's rain forest study served as the prototype for the big ecology that emerged from the IBP.[8]

The rain forest project demonstrated exactly how a scientific task force might go about studying a complex ecosystem. The work began with descriptive studies of forest structure. Scientists studied the ecological relations of some important species and analyzed patterns of material cycling and energy flow. Small areas of the forest were experimentally disturbed by clear-cutting and application of herbicides. More important, a radiation source (^{137}Ce) was used to irradiate a circular area of vegetation approximately 160 meters in diameter for a period of approximately three months. As discussed in chapter 6, this was one of several such radiation studies sponsored by the AEC in an effort to determine the ability of ecosystems to recover from the effects of radiation.[9] Finally, Odum used data collected from all phases of the task force to construct circuit models of ecosystem function.

Both the scope of the rain forest project and its implementation were impressive. It provided more concrete results than most later IBP projects, and it served as a benchmark for later large-scale studies of tropical rain forest ecosystems.[10] John Wolfe, director of the Environmental Sciences Division at the AEC, considered it one of the best scientific projects funded by the agency.[11] But the El Verde study also illustrated the problems of big ecology, problems that later plagued the IBP. The final product of the study was a massive ten-pound book, some 1,600 pages long. Odum's concluding chapter was "turgid with ideas," and to a certain extent it synthesized the results of the numerous individual investigations.[12] But the 111 chapters, written by various members of the task force, were rather uncoordinated. Critics complained that the authors sometimes contradicted one another, and related information was often scattered among several chapters. Although individual chapters contained much useful information, even scientists familiar with radiation biology and tropical ecology found it difficult to glean coherent conclusions from the book or to compare the overall results with those from other AEC studies.

Odum's claims about teamwork and planning, notwithstanding, the results of the El Verde study failed to form a unified whole. In sharp contrast to his influential studies of Silver Springs and Eniwetok Atoll, studies that could be easily read and understood by any ecologist, the rain forest book was too diffuse and disjointed. Plenty of data had been collected at El Verde, but as a reviewer pointed out, if one wanted to understand Odum's ideas about ecosystems, his slim book, *Environment, Power, and Society* was a better place to look.[13] This problem plagued most later IBP projects as well. It turned out to be far easier to recruit and support an ecological task force than to produce coherent results that other ecologists could easily use.

This type of brute force approach to research had other problems as well. Despite the large size of the book, Odum presented it as a preliminary report rather than a definitive study of a tropical rain forest. If Odum, however, envisioned his book as a point of departure for future research, there was no apparent mechanism for continuing big ecology at El Verde. The rain forest study had been only one of Odum's interests, and given his restless imagination he showed little inclination to make it his life's work. Many investigators on the project were graduate students whose ties to the enterprise had been limited to a year or two. Other members of the task force, biologists such as Carl Jordan, became leading authorities on tropical ecosystems, but they did so by developing independent research programs in areas other than El Verde. The El Verde site continued to be studied by scientists, but there was no institutional commitment to maintaining the type of large ecological task force that Odum and the AEC had gathered. Of course, scientific projects are nearly always incomplete, and the best pieces of research are often those that leave questions partially unanswered or point to interesting new problems. Nonetheless, given Odum's philosophy of big science, the overall value of the El Verde project can certainly be questioned. In Odum's scheme the ultimate goal of an ecological task force was to produce models that engineers could use to manage ecosystems. Such sophisticated and powerful models were not a product of the El Verde project. The suggestive preliminary network models that Odum created served as a heuristic tool for developing many ideas about environmental management that he discussed in *Environment, Power and Society*. But the dream of producing models that could be used directly for management purposes, a dream shared by many IBP ecologists, was unfulfilled.[14]

Great scientists do not necessarily make good generals. Odum's intellectual strengths were his fertile imagination, his innovative approaches to ecological theory, and his bold experimental techniques. Organizing and coordinating the work of an army of researchers was apparently not his forté. The logistical problems so evident in the rain forest study were not simply the result of flawed leadership on the part of a brilliant, but idiosyncratic, scientist; they also raise a fundamental question about the need for scientific task forces to carry out basic research in ecology. Could the El Verde ecosystem have been studied more economically and completely by a decentralized system of small research groups and independent scientists? Such a program might not have attracted funding from the AEC, but critics raised precisely this type of question in relation to the IBP biome studies.

The Origin of the International Biological Program

Planning for the International Biological Program grew out of the much publicized International Geophysical Year (IGY), 1957–1958.[15]

Impressed by what had been accomplished through international cooperation among physical scientists, biologists, particularly in Europe, began discussing the possibility of a similar venture in the life sciences. Under the energetic leadership of C. H. Waddington, president of the International Union of Biological Sciences, the program began to take form in 1962. Prior to Waddington's tenure, discussions about the IBP had been diffuse, with various biologists jockeying to have their specialties become the central focus of the program. To his credit, Waddington avoided such parochial squabbles. Although he was a geneticist, Waddington was particularly interested in directing the program toward ecosystem ecology, a relatively modest specialty with global implications. By 1964, the conceptual organization of the project was established under the general theme, "The Biological Basis of Productivity and Human Welfare." Although small projects in human genetics and physiology remained part of the IBP, this project became, for all intents and purposes, an international study of ecosystems.

Enthusiasm for the IBP ran high in Europe and Canada, but in the United States the program had an inauspicious beginning.[16] Big ecology generated a certain amount of excitement within one segment of the discipline, but within the scientific community at large there was no groundswell of support for the IBP. Even among ecologists influential voices spoke out strongly against the program. Initially, the American program had little focused direction, a point that opponents routinely emphasized in their criticisms. And most important, obtaining adequate funding posed a serious problem that took several years to resolve. Eventually the United States mounted the most extensive and elaborate research projects in the IBP, but that followed a slow start. By the time American scientists entered the field, Canadian and European projects were already well established.

Supporters of the IBP have portrayed their critics as reactionaries: lone-wolf individualists opposed to cooperative research; self-satisfied, establishment biologists protecting professional turf; and old-fashioned naturalists blind to the intellectual promise of the ecosystem concept.[17] Such individuals existed, but opposition to the IBP was both more diverse and, in some cases, more profound than this list of characterizations suggests. Understanding the problems facing the establishment of the IBP in the United States requires a careful consideration of criticism from both outside and inside the discipline of ecology.

Perhaps not surprisingly, molecular biologists, always a chauvinistic group, were highly critical of increased funding for ecology. The idea of an expensive, long-term program aimed at understanding ecosystems was anathema to those who believed that "any organism bigger than *E.coli* serves only to confuse the issue."[18] One official at NSF reported that her colleagues in molecular biology stopped talking to her once she became involved with the IBP.[19] Such an extreme reaction from members of a powerful discipline may appear incongruous, but supporters of the IBP openly discussed the possibility of raising ecology to the status of molecular biology.[20] They argued that there was a pressing social need for understanding ecosystems, that progress in this area would be expensive (several million dollars for each ecosystem studied), and that it would require the recruitment and training of large numbers of ecologists. The fact that supporters of the IBP could command prominent public forums for making these appeals may have given even well-funded molecular biologists cause for concern.

Opposition to the IBP from outside ecology might be expected, and in the American biological community, so dominated by molecular biology, this opposition undoubtedly had an adverse effect on establishing the program.[21] But prominent ecologists also voiced criticisms. For some, the need for large-scale, integrated research programs in ecology was not compelling.[22] To a certain extent this may have reflected a lone-wolf mentality, an aversion to having individual research projects dictated by scientific administrators, but it also reflected more serious misgivings about big science. Critics pointed to a major difference between the administrative structure of the IBP and its model, the IGY. The IGY involved actual multinational collaboration in simultaneous data collection around the globe. By its very nature, this type of research required large-scale cooperative efforts by scientists in several countries.

But the IBP represented a much looser form of collaboration. Each nation carried out independent research programs united only by the general theme of biological productivity. Such research was already being done in the United States, and there was every indication that ecosystem studies would continue, with or without the IBP. Furthermore, observers pointed out that increasing the funding for ecosystem research would not necessarily increase the quality of results.[23] Indeed, to the extent that productive scientists were drawn away from research and into IBP administration, the quality of ecosystem studies might actually decline. If so, the IBP would turn into a costly exercise in mediocre science, one that would produce little of value and bring discredit on ecology. This concern, that the IBP was a boondoggle, was raised numerous times during the planning and implementation of the program.

Inevitably, much internal criticism of the IBP revolved around money. LaMont Cole, a Cornell University ecologist who first dubbed the program a boondoggle, worried that it would turn into a raid on the U.S. Treasury to support foreign scientists.[24] Other critics also voiced concerns that American dollars would be siphoned off to pay for foreign research projects. Such concerns were exacerbated by fears that funding for the IBP would be diverted from ongoing research programs. As a matter of fact this did not happen, but criticisms of this type are revealing for two reasons. First, they reflect the deep intellectual division within the discipline of ecology. Many prominent ecologists were convinced that ecosystem studies, particularly on a large scale, did not constitute the best part of ecology. Second, these fears illustrate just how precarious many ecologists viewed their relationship with federal funding agencies. By the mid-1960s the "Golden Age" of federal funding had come to a close.[25] With inflation and increasing numbers of scientists competing for limited funds, ecologists were beginning to feel the funding pinch. For many ecologists, the IBP was perceived not as a mechanism for generating new financial support, but as an ominous threat to already scarce resources.

Criticism notwithstanding, supporters of the IBP began formally planning for scientific activities in the United States in the spring of 1963. The Ecological Society of America (ESA) set up a study committee, chaired by Eugene Odum, to serve as a liaison with European biologists and make recommendations to the society. Six months later the National Academy of Sciences (NAS) appointed an ad hoc committee to consider prospects for American participation in the IBP. As the official organization representing the United States on international scientific matters, the NAS was the logical place for planning IBP activities. But historically the NAS had largely ignored ecology; during the early 1960s only two ecologists were members of the academy. Therefore, academy participation in the planning process did little to reassure ecologists skeptical about the project.[26]

Relations between the NAS and professional ecologists became particularly strained a year later when Roger Revelle was appointed chairman of the U.S. National Committee for the IBP in January 1965. Ecologists were outraged that the academy had named a nonecologist, indeed a nonbiologist, to head the American effort.[27] This episode ruffled the feathers of professional ecologists, but Revelle was an excellent choice for the position. He had played an important role in establishing federal funding of academic science at the Office of Naval Research after World War II, he had been in charge of the scientific surveys of Bikini Atoll during Operation Crossroads, he had built Scripps Institution of Oceanography into a major research center, and he had extensive experience in international scientific affairs. A self-described "scientific administrator" he had the abilities needed to get the IBP off the ground. As C. H. Waddington noted approvingly, Revelle was a "real thruster."[28] No ecologist in the United States could match Revelle's skills as an administrator and lobbyist. When congressional support for the IBP was sought, Revelle's leadership proved crucial.

Appointing a national committee to oversee the planning for the IBP was only a beginning; it was three and a half years before the first American IBP research program began, and more than a year after that before Congress authorized full federal funding of the program. Conceptually, the IBP remained rather nebulous. W. Frank Blair, who eventually headed the U.S. National Committee on the IBP, described the committee as intellectually "floundering" during its first year of existence.[29] The tendency, according to Blair, was to promise something to any scientist who wanted to identify with the IBP. This situation changed quite dramatically after a meeting held in Williamstown, Massachusetts, in October 1966. Largely under the influence of the University of Michigan ecologist Frederick Smith, the committee agreed to orient the IBP around six biome studies, coordinated under an Analysis of Ecosystems Program. Smaller research programs, including one in evolutionary ecology were supported, but the lion's share of funding went to the ecosystem program. Shortly after the Williamstown meeting, the Analysis of Ecosystems Program, under the direction of Smith, obtained a small planning grant from the National Science Foundation (NSF) to begin organizing the biome studies.

Finding funding for actually carrying out the biome studies was a much greater problem. As a Washington insider, Revelle was instrumental in getting the process started. He was a personal friend of California Congressman George Miller, chairman of the House Committee on Science and Astronautics. Through Revelle's efforts, Miller agreed to introduce the first of several bills calling for federal recognition and support of the IBP. During the first round of subcommittee hearings Revelle provided eloquent testimony in favor of the program, but shortly thereafter he resigned as chairman of the U.S. National Committee on the IBP. He was replaced by Blair, a former president of the ESA and one of the earliest proponents of the IPB.

In his memoir, Blair recalls that the political climate in Washington was favorable to the IBP. By the mid-1960s, popular environmental concerns were beginning to influence policy makers, and the language of ecology may have struck a responsive chord in the capitol. In his critical history of the IBP, Chunglin Kwa, agreeing with Blair, argues that the metaphor of the self-governing machine, a metaphor shared by systems ecologists and the social engineers of the Great Society, facilitated the debate over the IBP.[30] If Blair and Kwa are correct, then the difficulty of obtaining adequate funding for the program is particularly striking. The first two IBP bills introduced in the House died in committee. After considerable debate an appropriations bill was finally passed by the House in November 1969. The companion bill in the Senate was not passed until nine months later. The legislation was finally signed into law by Richard Nixon on October 7, 1970, almost seven years after the National Academy of Sciences first began considering American participation in the IBP.

That ecologists eventually obtained some $50 million in support from the government was owing more to their dogged determination than to fortunate circumstances. Some of the most influential leaders of the IBP were openly contemptuous of the political process on which they so depended for financial support,[31] and, with the exception of Revelle, they were political neophytes. Congressional testimony by ecologists was often uncoordinated and ineffective. Indeed, when it came to testifying before congressional committees, supporters of the IBP sometimes seemed to be their own worst enemies.[32] With little political acumen or experience, supporters of the IBP had to learn the fine art of lobbying as they went along. Support for the program, even from supposed allies, was often lukewarm.[33] Harve Carlson, head of the Division of Biological Sciences at NSF, was sympathetic to the IBP, but within NSF and especially within the National Academy of Sciences skepticism about the program continued. Molecular biologists and other laboratory scientists actively lobbied against big ecology. Finally, despite concerns about pollution, public support for environmentalism was not particularly deep during the 1960s. The Clean Air Acts (1963, 1970) and the Water Quality Act (1965) attacked highly visible examples of pollution. These notable legislative achievements notwithstanding, during the 1960s there was no powerful constituency for whom the environment was a burning issue. Environmentalism was often viewed as a fringe political movement, and during the late 1960s grassroots activism was directed much more toward civil rights and the war in Vietnam. Compared to the "war on cancer," an emotional issue that brought tremendous political pressure to bear on public officials during this period, environmentalism could be safely ignored. Although the Johnson administration initially supported the IBP during early hearings on the program, it later reversed its position for budgetary reasons without worrying about the political consequences.[34]

Implementing the Biome Programs

Ecological research under the IBP consisted primarily of five large terrestrial biome studies, although a number of smaller projects were also supported.[35] The biome studies shared several important characteristics. In keeping with the overall IPB theme, all were oriented toward studying the structure and function of ecosystems. To a greater or lesser extent, all were conceived as big science projects. The National Science Foundation provided large block grants to each project; the project director then contracted scientists to complete the various parts of the study. Participating scientist were to make small, individual contributions to the larger study. But in some cases, at least, directors made no attempt to involve these scientists in the overall planning of the project. For many leaders of the IBP this was the essence of big ecology: implementing a hierarchical management approach to organize large teams of ecologists to attack problems of great complexity.[36] Finally, the ultimate goal of all these studies was the production of large systems models, as tools for both understanding and managing ecosystems. As Kwa has suggested, the organizational structure of the biome research teams mirrored the models of nature that these teams attempted to create.[37] In both cases, the principles of systems analysis were applied to large, complex systems.

The biome studies differed considerably in size and to the degree that they approached these ideals. Two studies, the Grassland Biome project and the Eastern Deciduous Forest Biome project, consumed nearly half the IBP budget for research.[38] Perhaps not surprisingly, both these projects had strong ties with the Oak Ridge National Laboratory, where scientists had fostered the philosophy of big science for a quarter of a century. Stanley Auerbach, who had started ecological research at ORNL in 1954, headed the forest study, and nearly 25 percent of the 176 participating scientists were recruited from that laboratory. Auerbach, a strong proponent of ecological task forces, believed it absolutely essential for ecologists to learn to work in large groups. This required these biologists to abandon their traditional individualistic attitudes and accept a new psychology and sociology of team research.[39] The organization of the biome study prevented Auerbach from fully implementing his managerial model. The Eastern Deciduous Forest Biome project was carried out at several different sites. A number of prominent ecologists worked on the project, some of whom were less than enthusiastic about having their research dictated from above. Auerbach ended up delegating much of his authority, and critics later charged that much research done on the forest study was not very different from the little science approach of traditional ecology.[40]

To a much greater degree the Grassland Biome project exemplified the characteristics of big science.[41] The study was directed by George Van Dyne, one of the original systems ecologists who had gathered at the ORNL during the early 1960s. Van Dyne was thirty-five years old when he was appointed director of the grassland biome study. Frederick Smith, who chose him to head the project, was impressed with not only Van Dyne's abilities as a systems modeler but also his encyclopedic knowledge of the prairie. An extraordinarily ambitious scientist, he had completed more than one hundred professional papers and written or edited nine books by the time of his death at age forty-eight. The team of researchers that Van Dyne began to organize in 1968 was reminiscent of Howard Odum's idea of an ecological task force. In both cases, researchers were organized hierarchically with systems ecologists playing the dominant role. And in both cases, the entire operation was tightly controlled by a charismatic leader who inspired loyalty and enthusiasm among subordinates. In the case of the Grassland Biome project, large numbers of workers, many of them graduate students, were hired to gather data primarily at a site near the Colorado State University campus.[42] A team of twenty-four systems ecologists used vast amounts of information collected on productivity, decomposition, nutrient cycling, hydrology, and weather to build a single model of the operation of the entire ecosystem.

Within the project, individual scientists had limited autonomy. By the end of the project several hundred articles had been published by members of the team, and more than one hundred graduate students had completed theses or dissertations related to the biome study.[43] Individual research efforts, however, were clearly subordinated to model building. Even defenders of the IBP admitted that many of Van Dyne's workers were mediocre scientists. One NSF official estimated that less than half could have won foundation grants on their own merits.[44] The justification for this was that large systems models required vast amounts of data. Given proper leadership, these data could presumably be collected by second-rate scientists. The total ecosystem model was to be more than the sum of these individual parts.

For true believers, building large systems models was the heart and soul of the IBP; the program sought to be the "proving ground" for systems ecology.[45] But even during the initial planning of the program there had been considerable skepticism toward this approach, and as the operational phase of the IBP unfolded enthusiasm for total systems models further eroded. In fact, modelers in the deciduous forest biome study soon abandoned plans for an overall systems model; instead they put their efforts into building smaller models of various parts of the ecosystem. The smaller biome studies also followed this modeling strategy. By the end of program, the grassland study was the only biome project that remained true to the IBP ideal of constructing a total systems model.

Most postmortem analyses of the IBP have not been kind to the grassland modeling effort. Chunglin Kwa has argued that members of the modeling team had important philosophical differences. George Van Dyne, viewing systems models primarily as a kind of heuristic tool, favored general models that could be used as a means for understanding ecosystems as complete functional entities. But his director of systems analysis, George Innis, had a much more restricted conception of systems ecology. For Innis, who designed the model, systems models were built to answer a small set of management-oriented questions concerning the effects of perturbations on ecosystems. As a mathematician he was willing to make considerable sacrifices in realism for gains in predictive power. Biotic interactions were largely ignored; the various possible states of the ecosystem were determined almost exclusively by abiotic factors. The possibility of inherent instabilities within ecosystems was also swept aside. The model was based on deterministic relationships between variables; a change in a controlling variable would lead to specific changes in other variables of the system. Despite these simplifying assumptions, the model was enormously complicated. Although its proponents claimed that the model provided important insights, it could only be used by those thoroughly familiar with it. Using a compelling evolutionary metaphor, Kwa describes the grassland model as a kind of computational giant Irish elk. Like the monstrously large antlers of the extinct elk, the sheer complexity of the total systems model proved maladaptive.

One justification for building systems models is their heuristic function. By formalizing what is known about an ecosystem, the model builder can identify what is not known about the system.[46] Critics charged, however, that the very complexity of total systems models led to ambiguity and actually detracted from their heuristic value. "Models are useful when they demonstrate clearly the implications of assumptions," wrote Daniel Botkin, "but the implications of manipulating a great many vaguely based parameters cannot be clear."[47] Despite the enormous amount of work involved in the grassland study,

Botkin argued that there was too little data and too little understanding of basic ecological processes to adequately model the whole ecosystem. As an alternative to such total system models, Botkin favored the use of smaller simulations directed at one or a few important ecological processes. His recommendation may have been self-serving, for this was precisely the type of model that Botkin had been building during the early 1970s. Nonetheless, the history of post-IBP modeling largely supported his contention that smaller models are more useful than the large-scale modeling effort of the Grassland Biome project.

Some modelers associated with the IBP seemed to agree with Botkin's criticism. A whimsical cartoon drawn by a researcher at Oak Ridge National Laboratory shows an ecologist leaping from top to top of a series of increasingly ornate pedestals representing the various stages of research associated with the IBP, culminating in the total system model.[48] From the system model, a grand pedestal festooned with trumpeters, the ecologist leaps toward a cloud, only to come crashing back to earth. Beneath the humor, the implied criticism was difficult to miss: the real-world application of total system models was an illusion.

The Legacy of the IBP

Given the fact that both supporters and critics closely identified the program with total ecosystem modeling, it is perhaps not surprising that the IBP ended with a somewhat tarnished image. There were many outspoken critics, and relatively few supporters claimed that the program had been an outstanding success. Criticism aside, ecology certainly enjoyed some lasting benefits from the IBP.[49] The international program fostered increased communication among American ecologists and European workers. Many graduate students were trained under IBP programs. Despite the failure to create a workable model of an entire ecosystem, systems ecology introduced a generation of ecologists to the use of computer simulation. Interest in ecological modeling, on a smaller scale, was undoubtedly stimulated by the IBP.[50] Most important, the IBP resulted in a quantum increase in funding for ecological research, funding that continued at approximately the same level after 1974.[51] Critics charged that this increase came at the expense of ecologists in other specialties.[52] There is an element of truth in such complaints, for the IBP was primarily an ecosystem research program. However, the money for IBP was not diverted from other established funding programs, and ecologists from other specialties, including evolutionary ecology, did play a minor role in some biome projects. Furthermore, there is little reason to believe that funding for ecological research of any kind would have increased significantly without the IBP. Indeed, when taking inflation into account, NSF support for ecological research was actually declining in the years immediately prior to the IBP.

These contributions to professional ecology were not insubstantial, yet it is difficult to escape the conclusion that the IBP failed to meet the great expectations of its supporters. Intellectually, the legacy of the IBP suffers in comparison with the IGY on which it was patterned. For example, the rise of plate tectonics and the consequent vindication of continental drift, which revolutionized the way geologists thought about the earth, were direct outcomes of the IGY. In contrast, it is fair to say that the IBP produced no such revolutionary change in the way ecologists thought about ecosystems. The final report on the IBP submitted to the NSF in 1975 concluded that "no major breakthrough in ecological theory or biome level perspectives has resulted [from the IBP] to date."[53] Supporters of the IBP challenged this conclusion. Urging a "wait-and-see" attitude, they claimed that the success of the program should be judged on the contributions of large synthesis volumes to be published in the years following the end of the IBP.[54] Many of these volumes did appear, but a decade after the program ended, few IBP studies had been distilled into "textbook ecology." Some textbook authors, ignoring the program altogether, discussed biomes without any reference to IBP projects. Even Eugene Odum paid little attention to it in his later textbooks.[55] Textbooks are not the only indicators of what is important in a scientific field, but the paucity of references to the program in leading textbooks is significant; it suggests that a diverse group of influential ecologists considered the IBP unworthy of detailed presentation to a generation of ecologists-to-be.

The IBP did much to consolidate ecosystem ecology as a scientific specialty. After 1974 the Ecosystem Studies Program at NSF, a direct descendent of the funding program for the IBP, continued to support ecosystem research. Its budget dwarfed those of other ecological programs at the foundation.[56] But this success came at considerable cost, for it exacerbated the intellectual schism within the discipline of ecology. Systems ecology, in the narrow sense of computer modeling, had never dominated ecosystems studies. But after the IBP, critics tended to equate the two, pointing to the failure of total ecosystem modeling as reflecting inherent flaws in ecosystem ecology.[57] They complained that ecosystem ecologists were only interested in constructing computer models for short-term predictions and that they ignored more general biological and evolutionary problems. From a broad historical perspective such criticisms had little basis in reality, for ecosystem ecology had always been more than systems modeling. Within the historical context of 1970s, however, these criticisms reinforced the distinction between evolutionary ecology and ecosystem ecology.

If the IBP had any effect on the image of ecosystem ecology, it was most certainly to equate the specialty with big science. Proponents of the program had actively promoted the view that ecosystem studies required large, hierarchically organized teams of researchers. This antagonized many ecologists. During the planning stages of the IBP, critics such as the University of Michigan ecologist Nelson Hairston voiced strong objections to big ecology. After the program ended, Hairston complained that big ecology had diverted money away from more worthy individual projects, and he bluntly dismissed the accomplishments of the IBP; he told a reporter that much of the research done in the biome projects was "pretty crappy stuff."[58] Critics continued to point to the insidious effect of big science on ecology, and they often equated this effect with ecosystem research, in general. Ecosystem ecologists had been seduced by money some critics complained.[59]

For critics, the IBP may have tainted ecosystem ecology with the odium of big science, but prior to the IBP most ecosystem studies had been shoestring operations. The classical ecosystem concept, a concept that remained largely unchanged by the events of the late 1960s and early 1970s, was almost entirely the product of individuals or small research groups. The early field studies of Raymond Lindeman, John Teal, Frank Golley, and Eugene Odum were all low-budget operations. Even Howard Odum's study of Silver Springs, by far the most ambitious of the early ecosystem studies, was organized as traditional small science. One might argue, of course, that by the mid-1960s ecosystem ecology had been pushed to the limits of small science. Further advances required a more structured team approach. Certainly this claim was made by ecologists such as Howard Odum, George Van Dyne, and Stanley Auerbach. By equating ecosystem ecology with the large-scale projects of the IBP, critics have at least tacitly agreed. Must we conclude that inevitably ecosystem ecology would evolve into the type of big science typified by the IBP? There is, I believe, strong evidence to reject this historical conclusion. During this era at least one impressive ecosystem study did not fit the IBP mold.[60] Unlike most biome studies, the Hubbard Brook ecosystem study produced results that quickly became textbook ecology. Although it required large numbers of researchers, the Hubbard Brook project was conceived on the informal, prebureaucratic model of traditional small science. This was hardly a shoestring operation, but in important ways it departed radically from Howard Odum's idea of an "ecological task force."

Hubbard Brook: An Alternative to Big Ecology

The Hubbard Brook ecosystem study started in 1963 with a modest three-year NSF grant of $59,400 to Gene E. Likens and F. Herbert Bormann (figure 10), both of whom were ecologists at Dartmouth College. Bormann, a botanist with a Ph.D. from Duke University, had been at Dartmouth since 1956. His early research had little to do with ecosystems, but when he visited the Hubbard Brook watersheds during the late 1950s he thought that it would be an ideal place to begin a large-scale study. After discussing this possibility with some of his colleagues, he turned his attention toward Hubbard Brook in a serious way in 1961, when a Dartmouth graduate student wrote a thesis on the geochemistry of iodine in the watershed. Likens, a limnologist who joined the biology department at Dartmouth shortly before the grant proposal was written, received his Ph.D. in zoology from the University of Wisconsin. The initial grant supported research on nutrient cycling in several watersheds within the Hubbard Brook Experimental Forest in New Hampshire.[61] During this early phase of the project, Bormann and Likens were joined by a small interdisciplinary group of scientists: Noye Johnson, a Dartmouth geochemist; Robert Pierce, a hydrologist with the U.S. Forest Service; and John Eaton, a research assistant in forest ecology at Dartmouth. After the initial grant, Bormann and Likens continued to be supported by a string of uninterrupted NSF grants. Although this support never rivaled that of the larger IBP biome projects, it was substantial. During the peak years from 1970 to 1976, annual support to Bormann and Likens never dropped below $180,000. Overall funding from NSF for the period between 1963 and 1976 was just less than $2 million. This amount includes neither grants awarded to other scientists working at the site nor support from the Forest Service, IBM Corporation, and other agencies and foundations involved in the program. By the late 1970s the project had evolved into a study of total ecosystem function, involving some fifty senior scientists and numerous graduate students. More than two hundred books and articles had been written about the Hubbard Brook ecosystem during this time.

Superficially, Hubbard Brook appears to be another example of the push toward big ecology, and some historians have presented it as such.[62] Certainly, it was much larger than ecosystem studies done prior to 1963, and in some ways the research was similar to that done in the larger biome programs.[63] Like the biome projects of the IBP, Hubbard Brook was a large-scale attempt to understand the ecosystem as a whole. Bormann and Likens were imbued with Eugene Odum's idea of a "new ecology" in which the ecosystem was the central focus of ecological study. They adopted Odum's holistic perspective: the parts of the ecosystem could only be understood within the context of the entire system.[64] Such understanding called for cooperative research by large numbers of scientists; it could not be accomplished by a handful of independent scientists.

Despite these similarities, there were significant differences between the Hubbard Brook study and other examples of big ecology. These differences were underscored by the fact that Hubbard Brook scientists remained somewhat aloof from the activities of the IBP.[65] Big science, as its advocates have always emphasized, is more than a matter of scale; it also has to do with the way research is organized. In sharp contrast to the military or corporate models favored by proponents of big science, Bormann and Likens adopted a kind of evolutionary model for research. Unlike Howard Odum's military metaphor of a task force methodically analyzing nature in a series of preestablished stages, the Hubbard Brook study gradually expanded during its first decade. The original focus on biogeochemistry served as an intellectual core around which other areas of research emerged. In contrast to the hierarchical structure of the task force model, Bormann and Likens favored a more individualistic approach to research. The small group of original researchers provided continuity to the project. As the project developed, this group gradually attracted other senior scientists, postdoctoral researchers, and graduate students. Bormann and Likens recruited scientists who they believed would contribute to the growth of the program as a whole. All team members were expected to share data with the rest of the group. In contrast to the hierarchical structure of a task force, however, individual scientists enjoyed much greater autonomy in developing individual lines of research. Camaraderie was encouraged by the communal living arrangement at the somewhat remote study site. "It is my feeling," Bormann recalled, "that almost all scientists who have worked at

Hubbard Brook look back on the experience as rewarding not only scientifically but also in terms of fun and quality of life."[66] Perhaps not surprisingly, Bormann and Likens had little difficulty attracting first-rate ecologists to join them at Hubbard Brook. In retrospect, the two founders cited this informal, nonbureaucratic approach to research as one of the most important factors in the success of the project.[67]

The organization of research was important, but Hubbard Brook also provided natural characteristics that contributed to the eventual success of the project.[68] The site encompassed seven small watersheds, relatively self-contained ecosystems. Because of their close proximity and similarity, some watersheds could be used as experimental systems and others as undisturbed controls. Large-scale experimentation, not a hallmark of many IBP biome studies, became a central focus of the Hubbard Brook study. For example, one watershed was completely deforested in 1965, and herbicides were applied for the following three years to suppress the growth of plants. The effects of this manipulation on nutrient leaching, biogeochemical cycling, and erosion were compared with an adjacent watershed that had been left in its natural state. The physical characteristics of the small watersheds also greatly facilitated the study of biogeochemical cycling. Geologically the watersheds were effectively watertight. Water entered the system in the form of rain and left through the small streams that drained each watershed. By placing concrete weirs at the base of each watershed investigators could monitor the nutrients leaving the system in stream water. Such weirs had long been used by forest scientists to study hydrology, but Hubbard Brook was the first site where these were used for biogeochemical studies. The approach worked so well that the weirs became ecological icons; pictures of the structures became almost as common in biology textbooks as the famous photograph of Watson and Crick in front of their double helix model.

The centerpiece of the Hubbard Brook study was Bormann and Likens's Biomass Accumulation Model. This conceptual model made general predictions about changes in hardwood forests following clear-cutting. In building this model, the two ecologists relied upon several sources of information. Studies of energy flow and nutrient cycling at the site could be used to make short-term predictions. However, to predict changes in a forest extending over decades and centuries, the two ecologists had to rely on less direct evidence. One source was a computer simulation of tree growth developed by a small team of modelers: Jim Janak, Daniel Botkin, and Jim Wallace. The JABOWA program, named after the three modelers, predicted biomass and nutrient changes in a forest under average growing conditions.

Because the short-term predictions made by the model corresponded reasonably well with data gathered from experimental sites, Bormann and Likens were confident in predictions extending over several hundred years. Computer simulation was important, but at Hubbard Brook formal systems ecology never gained the dominant role that it did in some IBP biome projects. Both Bormann and Likens were skeptical of "grand computerized models."[69] They used JABOWA, but their Biomass Accumulation Model also relied heavily upon classical ecosystem concepts, historical data, and the scientists's intuition about the nature of northern hardwood forests.

According to the Biomass Accumulation Model, a forest underwent four stages of development following clear-cutting.[70] During a period of *reorganization,* lasting one or two decades, the forest lost total biomass; net primary productivity declined, as did transpiration and nutrient uptake. At the same time export of nutrients from the system in stream flow markedly increased. In short, during this initial stage the ecosystem lost much of its ability to regulate the flow of energy and nutrients. During the *aggradation* phase, a period of a century or more, the system gradually regained its homeostatic capacity. Biomass steadily increased and reached a peak at the end of the phase. Export of nutrients from the ecosystem decreased and eventually came into balance with nutrient input. After the aggradation phase, total biomass declined slightly during a variable *transition* stage. The cause of this decline was the gradual death of old, massive trees in the ecosystem and their replacement by younger, smaller trees. During this stage the forest became a patchwork or mosaic of trees of various ages. This transition stage ended in the establishment of a *steady state,* with total biomass fluctuating around a mean. Such a state could be expected to occur three to five hundred years after the clear-cut.

Bormann and Likens's discussion of the steady state is intriguing, for it so vividly illustrates both the deep intellectual roots and the sophisticated development of ecosystem theory by the mid-1970s. Their idea of steady state drew heavily upon equilibrium ideas formulated during the early days of ecology by Stephen Forbes, Henry Chandler Cowles, and especially Frederic Clements. Indeed, in a statement that could hardly be considered fashionable during the 1980s, Bormann admitted that Clementsian ecology, with its heavy emphasis on process, provided him with a broad intellectual framework for explaining forest development at Hubbard Brook.[71] Equally important was the influence of post–World War II ecosystem theorists, particularly Eugene Odum. The Biomass Accumulation Model was, in fact, both an extension and a critique of the ideas that Odum had put forward in his controversial 1969 article, "The Strategy of Ecosystem Development."[72]

Odum's article can be thought of as an updated and greatly expanded version of Clementsian concepts. The developmental idea of succession was clearly stated in Odum's dichotomy between "immature" and "mature" systems. More important, Odum predicted several developmental trends during succession: biomass would increase, approaching an asymptote at system maturity; species diversity would increase, reaching a maximum at system maturity; ecosystem homeostasis would increase as diversity increased; and the ratio of photosynthesis to respiration would decrease, until the two processes were in approximate equilibrium at system maturity. Because Odum's paper made predictions that could be tested and because these predictions were presented as clear dichotomies between immature and mature ecosystems, it served as "a healthy stimulus" for the Hubbard Brook ecologists.[73] Bormann and Likens, however, presented evidence that contradicted some of Odums's predictions, and these differences were reflected in the Biomass Accumulation Model. According to Bormann and Likens, both biomass and species diversity reached a maximum during the aggradation phase of development and then declined as the ecosystem reached maturity. Diversity and ecosystem stability were not as closely related as Odum believed, a point that other ecologists were making on theoretical grounds. The relationship between photosynthesis and respiration was also more complicated than Odum had predicted. During the reorganization and transition phases, respiration actually exceeded photosynthesis. Thus, the ecosystem was operating at an energy deficit, drawing on energy reserves stored primarily in nonliving biomass in the soil.

By the time Bormann and Likens presented their Biomass Accumulation Model, equilibrium and climax concepts were coming under attack. Although the two ecologists were committed to the idea of a steady state, they were aware of its limitations. Ecosystems were homeostatic as Eugene Odum claimed, but Bormann and Likens also emphasized the important role of chance fluctuations and local disturbances in determining ecosystem structure and function. The steady state was not a stereotypical Clementsian climax; rather it was a "shifting-mosaic steady state" made up of irregular patches of vegetation of various ages.[74] Spatially heterogeneous, the ecosystem nevertheless exhibited some forms of equilibrium. Within the system, as a whole, photosynthesis and respiration could be expected to be more or less in balance, the total amount of biomass remained roughly the same, and the relative importance of various species remained fairly constant. This idealized picture of ecosystem structure and function drew heavily upon the Clementsian tradition in ecology, but by acknowledging that nature was a shifting mosaic, Bormann and Likens were also recognizing the resurgent tradition of Henry Allan Gleason.

Hubbard Brook was the outstanding example of the mature ecosystem studies that emerged during the 1970s. Like earlier studies it shared a focus upon energy flow, biogeochemical cycling, and succession. It drew heavily upon ideas associated with the traditional ecosystem concept: open system, process, development, homeostasis, and equilibrium. These concepts gave the study an intellectual coherence, but they were not used dogmatically. Bormann and Likens emphasized the complexity of ecological processes, and they acknowledged the important role that random events could play in the life of a forest. Odum's "Strategy of Ecosystem Development" had served an important function in generating hypotheses, but many predictions of his simple, linear model of succession turned out to be false. At Hubbard Brook, ecosystem development was not a strictly deterministic process leading to an easily definable climax. Rather, it was a constant battle between forces of development and diminishment, of order and chaos. Through the use of biogeochemical data, studies of community structure and function, natural history, and computer simulation, Bormann and Likens had constructed the most detailed and comprehensive interpretation of succession in a northern hardwood forest. The results of the study were significant theoretically, but they also had important implications for practical problems of watershed management, forest regeneration, and effects of acid rain.

On Being the Right Size

Was big ecology an inevitable stage in the development of ecosystem ecology? Certainly by the early 1970s the scope of research had greatly expanded. Individual scientists could no longer contemplate undertaking a comprehensive study of even a relatively small ecosystem such as a watershed at Hubbard Brook. Such studies required scores of specialists often working over the course of a decade or longer. Size is important, but the scale of research efforts is only one criterion of big science. Equally important, as proponents have pointed out, is the way this research is organized. During the post–World War II era big science has often meant hierarchically structured research teams based upon corporate or military models of organization. In their more sanguine moments, this is the characteristic of big ecology that proponents of the IBP extolled. It was, they claimed, the only way that complex ecological systems could be investigated. However, the experiences of ecologists during the 1960s and early 1970s raise serious questions about the wisdom of ecological task forces. In retrospect, the more loosely organized Hubbard Brook ecosystem study appears to have been much more effective than any IBP biome study. The Hubbard Brook study cost less than most of the biome studies, but it had a greater intellectual impact than all the IBP projects. Hubbard Brook was neither traditional small science nor big ecology; Bormann and Likens had struck a balance between the two extremes. From small beginnings, Bormann and Likens had created an impressive ecosystem study that was just the right size.

What makes for good science? The answer is elusive, but the Hubbard Brook Study suggests some likely factors for success. Hubbard Brook provided Bormann and Likens with ideal conditions for ecosystem research. The relatively small watersheds had well-defined boundaries. The experimental forest provided an area where scientists could establish a kind of natural laboratory for controlled experiments. Some watersheds could be manipulated and adjacent ones maintained as controls. The geological structure of the area allowed accurate measurements of nutrient cycling, and Bormann's innovative use of concrete weirs provided a new technique for making these measurements. The project began during a time when the federal funding of basic research was reaching its zenith. Beginning with a rather modest grant from the National Science Foundation, Bormann and Likens quickly established Hubbard Brook as a major research program that attracted a series of more substantial grants. Cooperation from the U.S. Forest Service was also crucial to the program's success. The Hubbard Brook Experimental Forest was a major government installation for studying hydrology, meteorology, and forest experimentation. The Forest Service was a source of much data used by Bormann and Likens. Robert Pierce, a Forest Service scientist, was an early member of the research team, and his supervision of the site played an important part in maintaining the cohesiveness of the program. Bormann and Likens seem to have had an intuitive sense of how to organize research along informal lines. Individuality and intellectual freedom were strongly encouraged in the work of outside investigators, but so were teamwork and cooperation. In this delicate balancing act there were always concerns that a "cancerous blow-up" might tear the project apart.[75] Friction may have developed among some individual investigators; however, relations appear to have been remarkably harmonious within this large group of scientists.

Bormann and Likens had a shrewd sense of how to present the results of their study to a broad audience, a key to the success of their program. Some critics of the IBP argued that in terms of numbers of publications per federal dollar spent, the Hubbard Brook project was more cost effective than any biome study.[76] Mere numbers can hardly be considered a useful indicator of success; more important indicators are where these publications appeared and how influential they turned out to be. Articles on Hubbard Brook appeared in a wide variety of specialist journals in ecology, geology, forestry, and related fields. A number of articles appeared in *Science*, *BioScience*, and other general scientific periodicals. Broad syntheses of a decade of research were presented in two important monographs. Some five thousand reprints of a 1970 *Scientific American* article on Hubbard Brook were sold within three years of publication.[77] Educational films, television programs, and audio tapes brought Hubbard Brook to the general public. So did public lectures; Bormann and Likens spoke frequently, not only to university audiences but also to numerous civic organizations. By the early 1970s Hubbard Brook was often used in college and high school textbooks as an important example of ecosystem research. The results of Hubbard Brook may have been intrinsically valuable, but this aggressive dissemination of results insured that the study would be widely recognized. By the end of the 1970s it was perhaps the best known comprehensive study of an ecosystem.

Epilogue
The Flights of Apollo

To my mind, the outstanding spin-off from space research is not new technology. The real bonus has been that for the first time in human history we have had a chance to look at the Earth from space, and the information gained . . . has given rise to a whole new set of questions and answers.

—J. E. LOVELOCK, *Gaia: A New Look at Life on Earth*

AS PREPARATIONS for the first Earth Day were nearing completion during the spring of 1970, a more dramatic event was unfolding in space. Two days into what seemed to be a routine journey to the moon, astronauts aboard Apollo 13 were suddenly faced with an emergency. Two hundred thousand miles from earth, an oxygen tank in the service module exploded. The accident placed the lives of three astronauts in jeopardy, but it also struck at the lifeblood of Apollo 13. Without oxygen the fuel cells that generated electricity for the space craft were inoperable. A simple mechanical failure in a single part had disrupted the operation of the whole system. America's third lunar landing would have to be abandoned.

The fateful accident, the tense three and a half days of maneuvering the crippled spacecraft, and the final dramatic return of the astronauts provided a cautionary lesson about the limits of technology. As a reporter for *TIME* commented, space travel had become so commonplace that a degree of hubris had developed both at NASA and in the minds of American citizens: "It [space travel] was not so much frail human flesh against the vast challenges of space as it was technicians remembering the sequence of switches to throw."[1] Apollo 13 dashed that sense of complacency. For Eugene Odum and Howard Odum, the incident also demonstrated the fundamental differences between artificial life-support systems and the life-support system of the biosphere.[2] The spacecraft was simply a storage system; sufficient food, water, and oxygen were carried on board for the mission, and human wastes were stored for later disposal. There was no way to replenish supplies or recycle wastes. A minor accident imperiled both the mission and the lives of the astronauts. In contrast, the biosphere is a regenerative system. The necessities of life are constantly recycled by plants, animals, and microorganisms. These processes of regeneration are controlled by complex self-regulatory or homeostatic mechanisms. The redundancy of these mechanisms buffers the entire biosphere from failures in individual parts of the system.

Both Odums had thought deeply about the nature of life-support systems. The third edition of Eugene Odum's *Fundamentals of Ecology*, which was in preparation during the Apollo 13 episode, included a chapter on space biology. Howard Odum was one of a number of ecologists who had experimented with artificial microcosms during the early 1960s. These "stripped down" ecosystems could be used to study ecological processes under controlled conditions by taking ecosystem studies into the laboratory, as it were. But the use of microcosms had a practical application, as well. NASA supported some of this research in an attempt to design self-sustaining life-support systems for space stations. Despite enthusiasm for the project, the feasibility of building really compact systems seemed doubtful. Truly self-sustaining artificial ecosystems would have to be very large; they needed to approximate the biological diversity of natural ecosystems and required the same intricate regulatory mechanisms found in nature. Smaller, less complex ecosystems always needed to be resupplied from earth; they could never sever the "umbilical cord" with the biosphere.[3]

If Apollo 13 could be used to contrast engineered systems and the biosphere, then it could also be used as a telling metaphor for the human condition. For several years before the unsuccessful mission, the idea of "spaceship Earth" had been used by environmentalists. The analogy seemed even more appropriate after the accident. Like Apollo 13, the earth could be thought of as a fragile package of life traveling through a hostile, nonliving environment. Unlike the spacecraft, however, a crippled earth had no safe haven to which it could return. Therefore, damaging the life-support systems of the biosphere was courting disaster. For Eugene Odum, the barren landscape surrounding the smelters at Copper Hill, Tennessee, provided an eerie suggestion of what the earth would be without its biosphere.[4] Just as the NASA engineers in Houston scrambled to piece together enough information about Apollo 13 and its problems to save the astronauts, Odum suggested that ecologists were in a race against time to understand the intricate machinery of the biosphere before it was destroyed.[5]

If the Apollo 13 episode suggested analogies between machines and the biosphere, then the space program had also provided another way of thinking about the Earth. Photographs taken from earlier Apollo flights showed dramatically the unique character of earth as a living planet. These images captured the imagination of the lay public and professional scientists alike. Eugene Odum included these photographs in the third edition of his *Fundamentals of Ecology,* and even today, a large, poster-size photograph of the deep blue globe covered with wispy white clouds dominates one wall of his office at the University of Georgia. This idea of a living planet found its most detailed literary expression in James Lovelock's 1979 book, *Gaia: A New Look at Life on Earth.*[6] Lovelock, an engineer, and Lynn Margulis, an evolutionary microbiologist, became the leading proponents of the view that the biosphere is controlled by living organisms. This "Gaia hypothesis" was not a new idea; something similar had been suggested nearly half a century earlier by V. I. Vernadsky. According to Vernadsky the unique composition of the earth's atmosphere was the product of biological processes. The idea that the chemistry of aquatic ecosystems is regulated by organic activities had also been briefly discussed by the biological oceanographer Alfred Redfield during the late 1950s.[7] But primarily Margulis's theoretical work on the evolution of eukaryotic cells gave this idea credibility. According to Margulis, the large amount of oxygen in the atmosphere is the product of photosynthesis, a process that evolved in certain bacteria some three billion years ago.[8] The introduction of oxygen into the atmosphere caused a fundamental change in the biogeochemistry of the earth and provided conditions favorable to the evolution of more complex eukaryotic cells. Life, so it was claimed, created the conditions that we now find on earth, and life continues to regulate these conditions through the complex feedback mechanisms of the world's ecosystems.

Lovelock's book was a popular success, but it was widely ignored by scientists. Naming his hypothesis after a Greek goddess was perhaps a poor strategy for catching the attention of professional biologists, and Lovelock repeatedly slipped into anthropomorphic and teleological language when discussing the biosphere. The idea that the earth is a kind of superorganism—worse still a thinking organism—led most scientists to dismiss the Gaia hypothesis as romantic fiction; yet the idea that biological processes play a dominant role in shaping the earth's chemistry and geology could not be so easily rejected. Stripped of its literary excesses, this weaker form of the hypothesis has become scientifically respectable, if still quite controversial.[9]

During the early 1980s when the Gaia hypothesis remained in the scientific wilderness, one of the few ecologists to discuss the idea in print was Eugene Odum. Although cautious toward Lovelock's claims, he was obviously attracted to the Gaia hypothesis.[10] As the philosophical leader of modern ecosystem ecology, Odum's interest in the hypothesis is not surprising. The idea that the biosphere is a homeostatic or cybernetic system of living and nonliving components was a central feature of the ecosystem concept that Odum did so much to create during the decades following World War II. Like Lovelock, Odum believed that homeostasis was more than a metaphor; it was a fundamental property characteristic of all biological systems. The biosphere was simply the most inclusive of these systems, and with global environmental problems such as the greenhouse effect it had become, in many ways, the most important. Although Lovelock rarely used the term ecosystem, many of his ideas meshed perfectly with Odum's broad systems approach to ecology. Indeed, in several ways Lovelock's popular book reflected both the strengths and the limitations of post–World War II ecosystem ecology.

The ecosystem is an intuitively appealing concept for most ecologists, even for those critical of the way ecosystem ecology has developed as a specialty. It is the only ecological concept that explicitly combines biotic and abiotic factors and places them on roughly equal footing. The nonliving environment is neither an inert stage on which biological dramas are acted out, nor is it a set of factors that imposes rigid restrictions on organisms. The relationship is more interactive than either alternative. Organisms not only adapt to the physical environment, but they also modify it. Population and community ecologists may recognize such interactions, but historically ecosystem ecologists most emphasized the importance of these complex relationships. If Margulis and Lovelock are correct about the biological origin of the earth's atmosphere, then this is simply the most dramatic example of a more general relationship between the living and nonliving worlds.

Central to the Gaia hypothesis is the role of microorganisms as the primary living agents of biogeochemical cycling. This reflects a more fundamental reorientation in modern biological thought. Traditionally, biologists have been categorized as botanists and zoologists, and even within specialties such as ecology it has been common to make distinctions between plant ecology and animal ecology. During the second half of the twentieth century this dichotomy has broken down as biologists recognized the existence of at least five kingdoms of living organisms. Ecosystem ecology, with its emphasis upon functional relationships, has played an important role in this intellectual shift. The idea of the ecosystem makes no sense without considering the important roles that certain bacteria play as producers in aquatic ecosystems, and the even more important roles that bacteria and fungi play as decomposers. This was implicit in Raymond Lindeman's trophic-dynamic paper, although he did not discuss microorganisms in detail. Beginning in the 1950s, at a time when almost all ecologists studied either plants or animals, Eugene Odum began calling for a reorientation in ecological research and teaching.[11] Citing the dearth of information on decomposition in ecosystems, Odum emphasized the "crying need" for microbial ecologists. Ecologists, in general, needed to set aside the traditional taxonomic distinction between plants and animals and replace it with a broader perspective that focused upon the functional roles a wide variety of organisms played within the ecosystem. Traditional college courses in ecology, which were almost always oriented toward zoology or botany, also needed to be replaced by broader ecosystem courses that included material on microorganisms, as well as plants and animals. Other prominent ecologists agreed. It is perhaps no coincidence that the major proponent of the modern five kingdom system, Robert Whittaker, was also involved in studying ecosystems.

Probably the most significant contribution of ecosystem ecology during the past fifty years has been to the environmental sciences. Even critics admit that the ecosystem is an inherently useful concept for discussing environmental problems. These problems are by their very nature ecosystem problems, involving the relationships between organisms and nonliving resources. Ecosystem ecology has a proud tradition of combining theory and application. The Hubbard Brook study is but one example of this. It was directed toward important theoretical problems in ecology such as the nature of succession and the structure of forest communities, but it also had a strong emphasis upon practical problems of forest management and the effects of acid rain on biological systems. Eugene Odum's *Fundamentals of Ecology* was unique among ecology textbooks in the degree to which it integrated pure and applied problems in ecology. The ecosystem concept provided Odum with a useful conceptual framework for discussing the role of ecology in solving pressing social problems. During the decade of the 1960s this was the dominant educational model for teaching ecology. Ironically, as the environmental movement began to gain momentum during the 1970s, Odum's older textbook began to be replaced by a newer generation of textbooks, many of which were written by population ecologists. None of these replacements had the strong emphasis upon environmentalism and social responsibility found in *Fundamentals of Ecology*.

Central to Lovelock's Gaia hypothesis are two related ideas: the biosphere is self-regulating, and this self-regulation is analogous to homeostatic mechanisms in organisms and cybernetic controls in automated machines. These themes weave through the modern history of ecology. G. Evelyn Hutchinson made the same claim about the biosphere almost half a century ago. Ecosystem ecologists, most notably the Odum brothers, generalized this idea to apply to all ecosystems. For Howard Odum, the parts of the ecosystem are regulated by complex information pathways, the invisible wires of nature. For Eugene Odum, information flows through the ecosystem much the way that hormones travel through the body. Through this complex mechanical or physiological process the stability of the system is maintained. Homeostasis or negative feedback do not imply perfect balance. Even in the most highly evolved homeostatic systems fluctuations and random disturbances occur. Nonetheless, the Odums's view of nature, with its emphasis upon self-regulation and stability, has been increasingly challenged by other ecologists. Critics have pointed to the often vague uses of stability, a term that can have several meanings in ecology.[12] They have challenged the idea that ecosystems are truly cybernetic and complained that ecosystem ecologists have misused a concept—negative feedback—borrowed from a different discipline.[13] On a more basic level, critics have attacked what they see as an overemphasis upon constancy, balance, and gradual change in traditional ecosystem ecology. In its place, they would erect a new ecology that emphasizes indeterminism, instability, and constant change. Ecosystems, so critics claim, may be perpetually out of balance.

The recent shift toward a nonequilibrium ecology seems quite dramatic. Proponents, who have often characterized themselves as a long embattled minority, now find themselves accused of creating a new orthodoxy.[14] Both characterizations are exaggerated. Take, for example, Daniel Botkin's *Discordant Harmonies*, perhaps the most detailed historical and philosophical statement of nonequilibrium ecology. Botkin repeats the well-worn claim that ecology is deeply rooted in the idea of a "balance of nature."[15] According to Botkin, this idea of harmonious balance, which is derived from both our Judeo-Christian heritage and, more important, machine age culture, pervades the early literature of ecology. Typical of this tradition is Stephen Forbes, the pioneer limnologist and ecologist, whom Botkin accuses of having believed that nature is in nearly perfect balance. Forbes did believe in equilibrium, but Botkin exaggerates Forbes's true position on the matter. A close reading of Forbes's work shows that, although he believed that nature maintained a high degree of constancy, he likewise acknowledged the importance of unpredictable variations in the environment, random fluctuations in populations, and catastrophic disturbances in communities.[16] Botkin's more indeterminate view of nature clearly differs from that of Forbes, but it is hardly the radical break with the past that Botkin claims. Botkin largely ignores more recent examples of equilibrium in the work of prominent ecologists such as G. Evelyn Hutchinson and the Odum brothers. Ironically, when Botkin suggests replacements for the misleading organic and mechanical metaphors used by ecologists in the past, he turns to computers—a metaphor also used by Howard Odum, one of the few ecologists for whom equilibrium really has been an idée fixe.

Closely related to the debate over the cybernetic nature of ecosystems is the problem of function. Like other ecosystem concepts the Gaia hypothesis is premised on the idea that organisms play functional roles in the biosphere. Indeed, Lovelock's blatant teleology highlighted this way of thinking about organisms and ecosystems. Supporters of the ecosystem concept argue that this type of teleological language can be avoided, but the more fundamental problem with functional thinking has rarely been addressed. In modern evolutionary ecology, the term *function* implies that an adaptation has evolved by natural selection to carry out a specific purpose. Most supposed ecosystem functions could only evolve through the discredited mechanism of group selection. How can ecosystem ecologists defend functionalism in the face of George Williams's critique of group selection? One approach has been suggested by David Sloan Wilson.[17] According to Wilson, if an individual's fitness is significantly determined by its interactions with other members of the community, then these interactions can properly be referred to as ecosystem functions. Wilson uses a restricted form of group selection to explain the evolution of these functions. This theoretical approach avoids many problems associated with the older group selection theories of V. C. Wynne-Edwards and Maxwell Dunbar, but it remains to be seen whether well-documented cases exist in nature. If Wilson could find such cases, his approach might reconcile evolutionary ecology and ecosystem ecology—a goal he clearly wishes to accomplish. At present, this goal remains a distant prospect.

The ramifications of the split that began during the mid-1960s between evolutionary ecology and ecosystem ecology continue. Among the strongest critics of Lovelock's Gaia hypothesis were evolutionary theorists who objected to the implications of group selection inherent in Lovelock's thinking. Ecosystem ecology, more generally, also continues to come under attack from critics for its lack of evolutionary emphasis. Ecosystem ecologists have been guilty of ignoring the modern synthesis. Consequently, when they do discuss evolution their comments often seem naive or muddled. But there is some validity in Eugene Odum's accusation that evolutionary ecologists ignore ecosystems. One insidious effect of the modern synthesis has been a narrowing of the focus of evolutionary ecology. Populations are important, but an evolutionary ecology worthy of the name must come to grips with the question of how large communities and ecosystems are structured. How this might be done remains unclear, for evolutionary and ecosystem ecologists have been talking past one another for almost a generation.

Whether or not a reunion occurs remains to be seen. An earlier generation of biologists was optimistic about creating a unified ecology, but their optimism was shattered by the group selection controversy of the 1960s. In retrospect, they recognized just how difficult this task might be.[18] Even conceptualizing the problem appears to be inherently difficult. Charles Darwin could be rigorously technical in his discussion of the evolution of adaptations, but when it came to discussing the complex interactions among organisms he often turned to a more literary style. During the late 1940s G. Evelyn Hutchinson hoped to create a formal mathematical synthesis of population ecology and biogeochemistry. The synthesis was never accomplished, and his students usually gravitated toward one area or the other. During the mid-1960s, as the group selection controversy was at its peak, Hutchinson published a series of essays entitled *The Ecological Theater and the Evolutionary Play*.[19] This evocative metaphor reflected the author's belief in the unity of science, but the essays provided few hints about where evolutionary ecologists and ecosystem ecologists might find common ground. Perhaps the two specialties have diverged so far that little common ground exists; however, a rapprochement would seem to benefit both specialties. Ecosystem ecology, despite its interdisciplinary trappings, is a biological science. Both historically and philosophically, its roots are in biology, not the physical sciences. Biology without a strong evolutionary core is misguided. But an evolutionary ecology that ignores ecosystems condemns itself to being a rather dowdy, academic discipline bereft of social or intellectual justification. Ecosystems may not be machines or organisms, but populations do interact with one another and with the surrounding physical environment. A century and a half ago, Darwin noted the importance of these interactions, writing that "plants and animals, most remote in the scale of nature, are bound together by a web of complex relations."[20] Today, in many parts of the earth this web has been badly damaged. Understanding the web, in all its complexity, is as central to evolutionary biology today as it was in 1859.

Notes

1. An Entangled Bank

1. Charles Darwin, *On the Origin of Species* [facsimile of the first edition, 1859] (Cambridge: Harvard University Press, 1964), 74, 489–490. This metaphor is discussed in quite a different context by Stanley Edgar Hyman, *The Tangled Bank: Darwin, Marx, Frazer, and Freud as Imaginative Writers* (New York: Atheneum, 1962), 32–34.
2. Darwin, *Origin of Species,* 73–74.
3. Donald Worster, *Nature's Economy: A History of Ecological Ideas* (Sierra Club Books, 1977; Cambridge: Cambridge University Press, 1985), chap. 8.
4. Ibid., 167.
5. Edward Manier, *The Young Darwin and His Cultural Circle* (Dordrecht: Reidel, 1978), 12. See also Michael Ruse, *The Darwinian Revolution: Science Red in Tooth and Claw* (Chicago: University of Chicago Press, 1979), 40–44, 55; Robert C. Stauffer, "Haeckel, Darwin, and Ecology," *Quarterly Review of Biology* 32 (1957): 138–144.
6. Darwin, *Origin of Species,* 62–63.
7. There is an extensive literature on the cultural impact of Darwinism. See Daniel Walker Howe, "Victorian Culture in America," in *Victorian America,* ed. Howe (Philadelphia: University of Pennsylvania Press, 1976); Robert C. Bannister, *Social Darwinism: Science and Myth in Anglo-American Social Thought* (Philadelphia: Temple University Press, 1979), chaps. 1–2; Paul F. Boller, Jr., *American Thought in Transition: The Impact of Evolutionary Naturalism, 1865–1900* (Chicago: Rand McNally, 1970); David W. Marcell, *Progress and Pragmatism: James, Dewey, Beard and the American Idea of Progress* (Westport: Greenwood Press, 1974), chaps. 1, 3; R. Jackson Wilson, *In Quest of Community: Social Philosophy in the United States, 1860–1920* (New York: John Wiley, 1968), chap. 1; John Higham, "The Reorientation of American Culture in the 1890s," in his *Writing American History: Essays on Modern Scholarship* (Bloomington: Indiana University Press, 1970); John Whiteclay Chambers II, *The Tyranny of Change: America in the Progressive Era, 1900–1917* (New York: St. Martin's Press, 1980), chap. 1.
8. Darwin, *Origin of Species,* 74–75. See also Bannister, *Social Darwinism,* chap. 1; Ruse, *The Darwinian Revolution,* 174–180, 238–239.

9. Daniel B. Botkin, *Discordant Harmonies: A New Ecology for the Twenty-first Century* (New York: Oxford University Press, 1990); Daniel Simberloff, "A Succession of Paradigms in Ecology: Essentialism to Materialism and Probabilism," *Synthese* 43 (1980): 3–39; John A. Wiens, "On Understanding a Non-Equilibrium World: Myth and Reality in Community Patterns and Processes," in *Ecological Communities: Conceptual Issues and the Evidence* ed. Donald R. Strong, Daniel Simberloff, Lawrence G. Abele, and Anne B. Thistle (Princeton: Princeton University Press, 1984). Several papers presented at the 1990 meeting of the Ecological Society of America at Snowbird, Utah, also expressed this point of view; for a review of these, see William W. Murdoch, "Equilibrium and Non-Equilibrium Paradigms," *Bulletin of the Ecological Society of America* 72, no. 1 (1991): 49–51.

10. Thomas Henry Huxley, "The Genealogy of Animals," in his *Critiques and Addresses* (New York: Appleton, 1873), 276.

11. The idea was a key element of German *Naturphilosophie.* See William Coleman, *Biology in the Nineteenth Century: Problems of Form, Function, and Transformation* (New York: John Wiley, 1971), 25–26. For a variety of twentieth-century expressions of this idea, see Charles Elton, *Animal Ecology and Evolution* (Oxford: Clarendon Press, 1930), 28; Paul Weiss, "The Problem of Specificity in Growth and Development," *Yale Journal of Biology and Medicine* 19 (1947): 235–278; F.J.R. Taylor, "Some Eco-evolutionary Aspects of Intracellular Symbiosis," *International Review of Cytology,* Supplement 14 (1983): 1–28; Lewis Thomas, *The Lives of a Cell: Notes of a Biology Watcher* (New York: Bantam Books, 1974), chap. 1.

12. Herbert Spencer, "The Social Organism," in his *The Man Versus the State,* ed. by Donald MacRae (*The Westminster Review,* January 1860, Baltimore: Penguin Books, 1969).

13. Bannister, *Social Darwinism,* chaps. 1–3. Richard Hofstadter, *Social Darwinism in American Thought,* 2d ed. (New York: George Braziller, 1955); Sidney Fine, *Laissez Faire and the General-Welfare State: A Study of Conflict in American Thought, 1865–1901* (Ann Arbor: University of Michigan Press, 1956), chap. 2; Cynthia Eagle Russett, *The Concept of Equilibrium in American Social Thought* (New Haven: Yale University Press, 1966), chap. 3; Robert Wiebe, *The Search for Order, 1877–1920* (Westport: Greenwood Press, 1967), chap. 6.

14. Worster, *Nature's Economy,* chap. 11; Ronald C. Tobey, *Saving the Prairies: The Life Cycle of the Founding School of American Plant Ecology, 1895–1955* (Berkeley: University of California Press, 1981), chap. 4; Robert P. McIntosh, *The Background of Ecology: Concept and Theory* (Cambridge: Cambridge University Press, 1985), 43; Sharon E. Kingsland, *Modeling Nature: Episodes in the History of Population Ecology* (Chicago: University of Chicago Press, 1985), chap. 1.

15. Russett, *Equilibrium,* chap. 3; Coleman, *Biology in Nineteenth Century,* chap. 5; John C. Greene, "Biology and Social Theory in the Nineteenth Century: Auguste Comte and Herbert Spencer," in *Critical Problems in the History of Science,* ed. Marshall Clagett (Madison: University of Wisconsin Press, 1969).

16. Spencer, "Social Organism," 200.

17. Ibid., 221–233.

18. J. Engelberg and L. L. Boyarsky, "The Noncybernetic Nature of Eco-

systems," *American Naturalist* 114 (1979): 317–324; Bernard C. Patten and Eugene P. Odum, "The Cybernetic Nature of Ecosystems," *American Naturalist* 118 (1981): 886–895.

19. Fine, *Laissez Faire,* 40.

20. Spencer, "Social Organism," 218.

21. John F. Stover, *American Railroads* (Chicago: University of Chicago Press, 1961), chap. 5; Wiebe, *Search for Order,* chap. 1; Chambers, *Tyranny of Change,* chaps. 3–4.

22. Alfred D. Chandler, Jr., *The Visible Hand: The Managerial Revolution in American Business* (Cambridge: Harvard University Press, 1977), chap. 4; Stover, *American Railroads,* chaps. 5–6. Robert Wiebe, *Search for Order,* characterizes 1876–1920 as a period of increasing bureaucratization and centralization in industrial America.

23. Thomas Henry Huxley, "Administrative Nihilism," in his *Critiques and Addresses,* 17–19.

24. Ibid., 19.

25. Charles Beard, *The Industrial Revolution,* 2d ed. (London: George Allen & Unwin, 1902), 90; see also Marcell, *Progress and Pragmatism,* chap. 7.

26. Stephen A. Forbes, "The Lake as a Microcosm," *Bulletin of the Illinois Natural History Survey* 15 (1925): 537–550 [originally published in the *Bulletin of the Peoria Scientific Association,* 1887]. McIntosh, *Background of Ecology,* 58–59; Kingsland, *Modeling Nature,* chap. 1; G. Evelyn Hutchinson, "*The Lacustrine Microcosm Reconsidered,*" *American Scientist* 52 (1964): 331–341.

27. Stephen Bocking, "Stephen Forbes, Jacob Reighard and the Emergence of Aquatic Ecology in the Great Lakes Region," *Journal of the History of Biology* 23 (1990): 461–498; Herbert B. Ward, "Stephen Alfred Forbes—A Tribute," *Science* 71 (1930): 378–381; Stephen A. Forbes, autobiographical letter, *Scientific Monthly* 30 (1930): 475–476; Harlow Mills, "Stephen Alfred Forbes," *Systematic Zoology* 13 (1964): 208–214; L. O. Howard, "Stephen Alfred Forbes, 1844–1930," *Biographical Memoirs of the National Academy of Sciences* 15 (1932): 1–54.

28. Forbes, autobiographical letter.

29. Ibid.

30. Forbes, "Microcosm," 549–550.

31. Ibid., 549.

32. Ibid., 539, 548.

33. Ibid., 550.

34. Ibid., 537. See also S. A. Forbes, "On Some Interactions of Organisms," *Bulletin of the Illinois State Laboratory of Natural History* 1 (1880): 3–17.

35. Forbes, "Microcosm," 547–548.

36. Ibid., 549–550.

37. Eugene Cittadino, "Ecology and the Professionalization of Botany in America, 1890–1905," *Studies in History of Biology* 4 (1980): 171–198. An excellent account of Bessey's role in shaping the development of ecology at the University of Nebraska can be found in Tobey, *Saving the Prairies,* chaps. 1–2.

38. Details of MacMillan's tenure at the University of Minnesota can be found in two unpublished histories in the Department of Botany Papers, University of Minnesota Archives: C. O. Rosendahl, "History of the Department," and Ernst C. Abbe, "An Informal History of the Department of Botany, University of Minnesota 1887–1950."

39. Conway MacMillan, *The Metaspermae of the Minnesota Valley: A List of the Higher Seed-Producing Plants Indigenous to the Drainage-Basin of the Minnesota River* (Minneapolis: Harrison & Smith, 1892).

40. Ibid., 583.

41. Ibid.

42. For a discussion of the important role that such *themata* play in scientific discovery, see Gerald Holton, *Thematic Origins of Scientific Thought: Kepler to Einstein* (Cambridge: Harvard University Press, 1973).

43. Botkin, *Discordant Harmonies,* 33, 41–43; Forbes, "Microcosm," 539; Forbes, "Interactions of Organisms," 8.

44. This possibility has received considerable attention from philosophers. An excellent, general discussion of the role of metaphors in science is provided by Mary B. Hesse, *Models and Analogies in Science* (Notre Dame: University of Notre Dame Press, 1966).

45. Forbes, "Microcosm," 537.

46. Wiebe, *Search for Order,* xiii.

47. Joel B. Hagen, "Organism and Environment: Frederic Clements's Vision of a Unified Physiological Ecology," in *The American Development of Biology,* eds. Ronald Rainger, Keith R. Benson, and Jane Maienschein (Philadelphia: University of Pennsylvania Press, 1988).

48. Forbes, "Interactions of Organisms," 3–4.

49. Peter J. Taylor, "Technocratic Optimism, H. T. Odum, and the Partial Transformation of Ecological Metaphor after World War II," *Journal of the History of Biology* 21 (1988): 213–244.

50. Strong, et al., *Ecological Communities,* vii.

2. A Rational Field Physiology

1. Frederic Edward Clements, *Research Methods in Ecology* (Lincoln: University Printing Company, 1905), 7.

2. Ibid., 10.

3. Eugene Cittadino, "Ecology and the Professionalization of Botany in America, 1890–1905," *Studies in History of Biology* 4 (1980): 171–198; Ronald Tobey, "Theoretical Science and Technology in American Ecology," *Technology and Culture* 17 (1976): 718–728; Richard A. Overfield, "Charles Bessey: The Impact of the 'New' Botany on American Agriculture, 1880–1910," *Technology and Culture* 16 (1975): 162–181.

4. Lawrence R. Veysey, *The Emergence of the American University* (Chicago: University of Chicago Press, 1965); Burton J. Bledstein, *The Culture of Professionalism: The Middle Class and the Development of Higher Education in America* (New York: Norton, 1976), chaps. 6–8.

5. Philip J. Pauly, "The Appearance of Academic Biology in Late Nineteenth-century America," *Journal of the History of Biology* 17 (1984): 369–397; Ronald C. Tobey, *Saving the Prairies: The Life Cycle of the Founding School of American Plant Ecology, 1895–1955* (Berkeley: University of California Press, 1981), chaps. 1–2; Jane Maienschein, "Whitman at Chicago: Establishing a Chicago Style of Biology?" in *The American Development of Biology,* ed. Ronald Rainger, Keith R. Benson, and Jane Maienschein (Philadelphia: University of

Pennsylvania Press, 1988); J. Ronald Engel, *Sacred Sands: The Struggle for Community in the Indiana Dunes* (Middletown: Wesleyan University Press, 1983), chaps. 2, 4.

6. Frederick Jackson Turner, "The Significance of the Frontier in American History," *Annual Report of the American Historical Association* (1893): 199–227. See also William Coleman, "Science and Symbol in the Turner Frontier Hypothesis," *American Historical Review* 72 (1966): 22–33.

7. Paul B. Sears, "Some Notes on the Ecology of Ecologists," *Scientific Monthly* 83 (1956): 22–27.

8. Stephen Pyne notes the "uncanny resemblance" between the developmental ideas used by early ecologists and Turner's frontier thesis; see *Fire in America: A Cultural History of Wildland and Rural Fire* (Princeton: Princeton University Press, 1982), 492. Engel, *Sacred Sands,* chap. 4, suggests a strong connection between the unique ecological and social theories developed at the University of Chicago.

9. Engel, *Sacred Sands,* chap. 2.

10. Pauly, "Appearance of Academic Biology"; Maienschein, "Whitman at Chicago"; Engel, *Sacred Sands,* chap. 4.

11. Henry Chandler Cowles, "The Ecological Relations of the Vegetation on the Sand Dunes of Lake Michigan," *Botanical Gazette* 27 (1899): 95–117, 167–202, 281–308, 361–391.

12. Cowles, "Ecological Relations," 97–98, 386; William Coleman, "Evolution into Ecology? The Strategy of Warming's Ecological Plant Geography," *Journal of the History of Biology* 19 (1986): 181–196.

13. Cowles, "Ecological Relations," 95.

14. Ibid., 194.

15. Ibid., 95, 194–195.

16. Ibid., 107.

17. Ibid., 381.

18. Ibid., 177.

19. Ibid., 302

20. Jerry S. Olson, "Rates of Succession and Soil Changes on Southern Lake Michigan Sand Dunes," *Botanical Gazette* 119 (1958): 125–170. This article won the prestigious Mercer Award from the Ecological Society of America.

21. Eugene P. Odum, *Fundamentals of Ecology,* 2d ed. (Philadelphia: Saunders, 1959), 257.

22. Engel, *Sacred Sands,* chap. 4. Prominent ecologists who studied under or were strongly influenced by Cowles early research include William S. Cooper, Paul Sears, Stanley Cain, and Victor Shelford.

23. Cowles, "Ecological Relations," 96, 194. For his comments on the chaotic state of ecological thought, see Henry Chandler Cowles, "The Work of the Year 1903 in Ecology," *Science* 19 (1904): 879–885.

24. Tobey, *Saving the Prairies,* 45–47. A more general account of this pragmatic revolt in American thought can be found in David W. Marcell, *Progress and Pragmatism: James, Dewey, Beard and the American Idea of Progress* (Westport: Greenwood Press, 1974), chap. 1.

25. Overfield, "Charles E. Bessey"; Tobey, "Theoretical Science"; Tobey, *Saving the Prairies,* chap. 2.

26. For example, see F. E. Clements to A. G. Tansley, May 7, 1918, Edith S. and Frederic E. Clements Collection, American Heritage Center, University of Wyoming, Box 62; henceforth, CC. See also Ernst C. Abbe, "An Informal History of the Department of Botany, University of Minnesota, 1887–1950," Department of Botany Papers, University of Minnesota Archives.

27. Roscoe Pound and Frederic E. Clements, *The Phytogeography of Nebraska,* 2d ed. (Lincoln: Botanical Seminar, 1900).

28. Clements, *Research Methods,* 199–200; Frederic E. Clements, *Plant Succession: An Analysis of the Development of Vegetation* (Washington, D.C.: Carnegie Institution of Washington, 1916), 3; Frederic E. Clements, John E. Weaver, and Herbert C. Hanson, *Plant Competition: An Analysis of Community Functions* (Washington, D.C.: Carnegie Institution of Washington, 1929), 314–318. See also Joel B. Hagen, "Organism and Environment: Frederic Clements's Vision of a Unified Physiological Ecology," in *American Development of Biology,* ed. Rainger, Benson, and Maienschein.

29. David Marcell, *Progress and Pragmatism,* 37. Numerous examples of organismal metaphors and analogies are discussed in Richard Hofstadter, *Social Darwinism in American Thought,* 2d ed. (New York: George Braziller, 1955), and Cynthia Eagle Russett, *The Concept of Equilibrium in American Social Thought* (New Haven: Yale University Press, 1966).

30. C[harles]. R[eid]. B[arnes]., "Physiology and Ecology," *Botanical Gazette* 44 (1907): 307–309; F. F. Blackman to A. G. Tansley, undated, A. G. Tansley Papers, Department of Botany, Cambridge University; Janice Emily Bowers, *A Sense of Place: The Life and Work of Forrest Shreve* (Tucson: University of Arizona Press, 1988), 59.

31. Clements, *Research Methods,* 129–144.

32. Ibid., 133.

33. Joel B. Hagen, "Experimentalists and Naturalists in Twentieth-Century Botany: Experimental Taxonomy, 1920–1950," *Journal of the History of Biology* 17 (1984): 249–270.

34. Clements, *Research Methods,* 239–241; Clements, *Plant Succession,* chap. 1.

35. Frederic E. Clements, *Plant Physiology and Ecology* (New York: Henry Holt, 1907), 252–253. Clements, Weaver, and Hanson, *Plant Competition,* 10–12.

36. Clements, *Research Methods,* 286.

37. Clements, *Plant Succession,* 80.

38. Ibid., 125, 145.

39. Ibid., 99.

40. Engel, *Sacred Sands,* chap. 4.

41. Henry Chandler Cowles, "The Physiographic Ecology of Chicago and Vicinity; A Study of the Origin, Development, and Classification of Plant Societies," *Botanical Gazette* 31 (1901): 73–182, 81.

42. Clements, *Plant Succession,* 3, 100.

43. Ibid., 99.

44. F. E. Clements, "The Life History of Lodgepole Burn Forests," *United States Forest Service Bulletin,* no. 79 (1910).

45. Malcolm Nicholson, "Henry Allan Gleason and the Individualistic Hypothesis: The Structure of a Botanist's Career," *Botanical Review* 56 (1990): 91–161; Robert P. McIntosh, "H. A. Gleason—'Individualistic Ecologist'

1882–1975: His Contributions to Ecological Theory," *Bulletin of the Torrey Botanical Club* 102 (1975): 253–273; Bassett Maguire, "Henry Allan Gleason," *Bulletin of the Torrey Botanical Club* 102 (1975): 274–282.

46. Henry Allan Gleason, "The Structure and Development of the Plant Association," *Bulletin of the Torrey Botanical Club* 44 (1917): 463–481, 464. Gleason's concept was refined in two later articles: "The Individualistic Concept of the Plant Association," *Bulletin of the Torrey Botanical Club* 53 (1926): 7–26; and "The Individualistic Concept of the Plant Association," *American Midland Naturalist* 21 (1939): 92–110. For a detailed discussion of the development of Gleason's argument, see Nicholson, "Gleason and the Individualistic Hypothesis."

47. Gleason, "Individualistic Concept (1926)," 9.

48. Ibid., 26.

49. H. A. Gleason, autobiographical letter, *Bulletin of the Ecological Society of America* 34 (1953): 40–42; McIntosh, "H. A. Gleason"; Robert Whittaker, "An Hypothesis Rejected: The Natural Distribution of Vegetation," in *Botany: An Ecological Approach,* ed. William A. Jensen and Frank B. Salisbury (Belmont: Wadsworth, 1972), pp. 689–691; Daniel Simberloff, "A Succession of Paradigms in Ecology: Essentialism to Materialism and Probabilism," *Synthese* 43 (1980): 3–39.

50. Clements characterized Gleason as "very likeable personally," but he considered Gleason's ecological research to be "unremarkable." See Clements to H. M. Hall, December 3, 1918, CC, Box 62.

51. Bowers, *Sense of Place,* chap. 8; Nicholson, "Gleason and the Individualistic Hypothesis."

52. Herbert Spencer, "The Social Organism," in his *The Man Versus the State,* ed. by Donald MacRae (Baltimore: Penguin Books, 1969); E. P. Odum, *Fundamentals of Ecology,* 2d ed., 246.

53. Raymond L. Lindeman Papers, Box 5, Sterling Library, Yale University.

54. Robert Harding Whittaker, "A Vegetation Analysis of the Great Smoky Mountains" (Ph.D. diss.: University of Illinois, 1948), 160.

55. Whittaker, "Hypothesis Rejected."

56. Robert H. Whittaker, *Communities and Ecosystems* (New York: MacMillan, 1970; 1975), 1–2, 294–296.

57. Robert H. Whittaker, "A Consideration of Climax Theory: The Climax as a Population and Pattern," *Ecological Monographs* 23 (1953): 41–78, 53.

58. Eugene P. Odum, *Fundamentals of Ecology,* 3rd ed. (Philadelphia: W. B. Saunders Co., 1971), 251. F. H. Bormann, "Lessons from Hubbard Brook," in *Proceedings of the Chaparral Ecosystems Research Meeting,* ed. E. Keller, S. Cooper, and J. DeVries, Report #62 of the California Water Resources Center (Davis: University of California, 1985).

3. An Ambiguous Legacy

1. T.F.H. Allen, "The Noble Art of Philosophical Ecology," *Ecology* 62 (1981): 870–871: Robert P. McIntosh, *The Background of Ecology: Concept and Theory* (Cambridge: Cambridge University Press, 1985), 80.

2. J. Braun-Blanquet, *Plant Sociology: The Study of Plant Communities*, trans. George D. Fuller and Henry S. Conard (New York: McGraw-Hill, 1932), 315; others quoted from A. G. Tansley, "The Classification of Vegetation and the Concept of Development," *Journal of Ecology* 8 (1920): 118–144.

3. McIntosh, *Background of Ecology*, 80.

4. Gerald L. Geison, "Scientific Change, Emerging Specialties, and Research Schools," *History of Science* 19 (1981): 20–40; Gerard Lemaine, Roy Macleod, Michael Mulkay, and Peter Weingart, eds., *Perspectives on the Emergence of Scientific Disciplines* (Chicago: Aldine, 1976); David L. Hull, *Science as a Process: An Evolutionary Account of the Social and Conceptual Development of Science* (Chicago: University of Chicago Press, 1988), chap. 10.

5. Nathan Reingold, "National Science Policy in a Private Foundation: The Carnegie Institution of Washington," in *The Organization of Knowledge in Modern America, 1860–1920*, ed. Alexandra Oleson and John Voss (Baltimore: Johns Hopkins University Press, 1979); Howard S. Miller, *Dollars for Research: Science and Its Patrons in Nineteenth-century America* (Seattle: University of Washington Press, 1970), chap. 9.

6. Reingold, "National Science Policy."

7. Janice Emily Bowers, *A Sense of Place: The Life and Work of Forrest Shreve* (Tucson: University of Arizona Press, 1988), chap 2.; Frederick Vernon Coville and Daniel Trembly MacDougal, *Desert Botanical Laboratory of the Carnegie Institution* (Washington, D.C.: Carnegie Institution of Washington, 1903); William G. McGinnies, *Discovering the Desert: Legacy of the Carnegie Desert Botanical Laboratory* (Tucson: University of Arizona Press, 1981).

8. Frederic E. Clements, Application for grant in aid of research and attached memorandum of Frederick Coville, Carnegie Institution of Washington Archives (File: Ecology: Director—Personnel); henceforth, CIW.

9. Ibid.

10. Edith S. Clements, *Adventures in Ecology: Half a Million Miles . . . From Mud to Macadam* (New York: Pageant Press, 1960), chaps. 5, 7; D. T. MacDougal to R. S. Woodward, January 12, 1914; Woodward to MacDougal, February 9, 1914, CIW (File: Botanical Research, Research Associates Part II); Minutes of the Executive Committee of the Carnegie Institution of Washington, February 18, 1915, CIW.

11. Reingold, "National Science Policy."

12. Robert S. Woodward, "Report of the President," *Carnegie Institution of Washington Yearbook* 15 (1916): 3–36.

13. Letter of Appointment to Frederic Clements, March 17, 1917, CIW (File: Ecology: Director—Personnel).

14. Frederic Clements to Robert S. Woodward, August 15, 1919; Woodward to Clements, August 20, 1919, Carnegie Institution of Washington Archives (File: Ecology: Projects Proposed).

15. Reingold, "National Science Policy."

16. Bowers, *Sense of Place*, 55.

17. During the summer of 1919 Clements was hospitalized for "nervous exhaustion"; see F. E. Clements to R. S. Woodward, June 2, 1919; Woodward to Clements, June 19, 1919, Dr. Edith S. Clements and Dr. Frederic E. Clements Collection, Box 62, American Heritage Center, University of Wyoming; henceforth, CC.

18. C. E. Bessey to C. O. Rosendahl, April 23, 1907; Francis Ramaley to Rosendahl, April 18, 1907; and Frederic Clements to Rosendahl, May 9, 1907; Department of Botany Papers, University of Minnesota Archives; henceforth, DBM.

19. Clements to Rosendahl, May 9, 1907, DBM. Some details of Clements's tenure at the university are provided in two unpublished histories of the department: Ernst C. Abbe, "An Informal History of the Department of Botany, University of Minnesota, 1887–1950"; and C. O. Rosendahl, "History of the Department," DBM.

20. J. C. M[erriam]., "Proposals Regarding Plan of Organization of Research in Plant Biology in the Carnegie Institution of Washington," unpublished memorandum, October 1, 1927, CIW (File: J. C. Merriam Memoranda: Plant Biology).

21. In 1951 this research unit was renamed the Department of Plant Biology, its current title.

22. J. C. M[erriam]., "Proposals."

23. Frederic Edward Clements, *Research Methods in Ecology* (Lincoln: University Publishing Company, 1905), 1.

24. Frederic Clements to J. C. Merriam, January 21, 1927; Merriam to Clements, January 31, 1927; Frederic E. Clements, "Development of Ecological Researches 1892–1927," unpublished memorandum (undated), CIW (File: Ecology: Miscellaneous).

25. C[harles]. R[eid]. B[arnes]., "Physiology and Ecology," *Botanical Gazette* 44 (1907): 307–309.

26. Bowers, *Sense of Place,* 59; emphasis in original text.

27. F. F. Blackman and A. G. Tansley, "Ecology in its Physiological and Phytotopographical Aspects," *The New Phytologist* 4 (1905): 199–203, 232–253.

28. F. F. Blackman to A. G. Tansley, undated, A. G. Tansley Papers, Department of Botany, Cambridge University; henceforth, TP.

29. Evidence for this can be found in extensive correspondence among Spoehr, Merriam, and Clements during the summer of 1929, CIW (File: Laboratory for Plant Physiology: Directors file).

30. Joel B. Hagen, "Experimentalists and Naturalists in Twentieth-Century Botany: Experimental Taxonomy, 1920–1950," *Journal of the History of Biology* 17 (1984): 249–270.

31. H. M. Hall, "Research on the Problem of Environmental Influence," unpublished memorandum to J. C. Merriam, July 31, 1925; E. B. Babcock to C. B. Hutchinson, August 12, 1925, CIW (File: Ecology: Miscellaneous).

32. A. F. Blakeslee to J. C. Merriam, December 7, 1924, CIW (File: Biological Conference 1925). See also C. B. Davenport to J. C. Merriam, December 5, 1924, in the same file.

33. Blakeslee to Merriam, December 7, 1924, CIW (File: Biological Conference 1925). For other evidence of the early cooperation between Clements and geneticists, see Blakeslee to Clements, April 4, 1923; Clements to Blakeslee, April 24, 1923; Clements to G. H. Shull, January 18, 1923, CC, Box 63.

34. F. E. Clements to J. C. Merriam, December 2, 1928, CIW (File: Plant Biology: Investigations of Dr. Clements, 1927–1929).

35. J. C. Merriam to H. A. Spoehr, December 8, 1928; Spoehr to Merriam, December 12, 1928, CIW (File: Plant Biology: Investigations of Dr. Clements, 1927–1929).

36. Clements developed an almost paranoid attitude toward the reception of his ideas by other biologists; he complained to the president of the Carnegie Institution that false rumors were being spread about the insignificance of his work. See F. E. Clements to J. C. Merriam, May 29, 1929; Clements's wife also complained bitterly of the existence of enemies within the institution who were undermining her husband's position; see Edith Clements to J. C. Merriam, August 17[?], 1929, CIW (File: Plant Biology: Investigations of Dr. Clements, 1927–1929).

37. Edith S. Clements to H. A. Spoehr, October 11, 1946; Spoehr to E. S. Clements, November 5, 1946, CIW (File: Clements, Frederic: Retirement). A general discussion of Clements's experiments and neo-Lamarckian conclusions, but with little actual experimental data, can be found in Frederic E. Clements, Emmett V. Martin, and Frances L. Long, *Adaptation and Origin in the Plant World: The Role of Environment in Evolution* (Waltham: Chronica Botanica, 1950). A more orthodox Darwinian interpretation of similar experiments can be found in Jens Clausen, David D. Keck, and William M. Hiesey, *Experimental Studies on the Nature of Species. I. Effect of Varied Environments on Western North American Plants* (Washington, D.C.: Carnegie Institution of Washington, 1940).

38. Hagen, "Experimentalists and Naturalists"; Lincoln Constance, *Botany at Berkeley: The First Hundred Years* (privately published, 1978).

39. Harvey Monroe Hall and Frederic E. Clements, *The Phylogenetic Method in Taxonomy: The North American Species of Artemesia, Chrysothamnus, and Atriplex* (Washington, D.C.: Carnegie Institution of Washington, 1923). I have discussed the controversy surrounding this book in my article, "Experimentalists and Naturalists."

40. Ernest Brown Babcock and Harvey Monroe Hall, "*Hemizonia congesta.* A Genetic, Ecologic, and Taxonomic Study of the Hay-Field Tarweeds," *University of California Publications in Botany* 13 (1924): 15–100.

41. Hall, "Research on Environmental Influence"; J. C. Merriam, "Proposals Regarding Plant of Organization of Research in Plant Biology in the Carnegie Institution," unpublished memorandum of J. C. Merriam, October 1, 1927, CIW (File: J. C. Merriam: Memoranda: Plant Biology).

42. Hagen, "Experimentalists and Naturalists."

43. F. E. Clements to J. C. Merriam, August 12, 1929; CIW (File: Plant Biology: Investigations of Dr. Clements, 1927–1929). Other botanical research programs, notably that of Forrest Shreve, also suffered under Spoehr's dictatorial administration; see Bowers, *Sense of Place,* chap. 17.

44. See the annual reports on ecological research in the *Carnegie Institution of Washington Yearbooks* between 1918 and 1941.

45. Clements clearly recognized that the reorganization of botanical research was a threat to his position within the institution. See Frederic E. Clements to J. C. Merriam, January 18, 1928; Edith Clements to J. C. Merriam (received January 23, 1928); Edith Clements to J. C. Merriam, August 17[?], 1929, CIW (File: Plant Biology: Investigations of Dr. Clements, 1927–1929).

46. R. S. Woodward to F. E. Clements, January 28, 1919, CC, Box 62.

47. E. Clements, *Adventures in Ecology,* 102.

48. This list includes Victor Shelford, Homer Shantz, Herbert Hanson, John Weaver, Charles Vorhies, Lee Dice, William Penfound, Arthur Vestal, and Stanley Cain—all of whom held elective offices in the Ecological Society of America.

49. F. E. Clements to J. E. Weaver, April 11, 1918, CC, Box 62.

50. These included *Plant Indicators: The Relation of Plant Communities to Process and Practice* (1920); *Aeration and Air-Content: The Role of Oxygen in Root Activity* (1921); *Experimental Pollination: An Outline of the Ecology of Flowers and Insects* (1923), coauthored by Frances Long; *The Phylogenetic Method in Taxonomy: The North American Species of Artemesia, Chrysothamnus, and Atriplex* (1923), coauthored by Harvey Hall; *Experimental Vegetation: The Relation of Climaxes to Climate* (1924), coauthored by John Weaver; *The Phytometer Method in Ecology: The Plant and Community as Instruments* (1924), coauthored by Glenn Goldsmith; and *Plant Competition: An Analysis of Community Functions* (1929), coauthored by John Weaver and Herbert Hanson.

51. For example, see Garland E. Allen's discussion of the group dynamics of Morgan's *Drosophila* research team in *Thomas Hunt Morgan: The Man and His Science* (Princeton: Princeton University Press, 1978), 188–213.

52. For a more detailed analysis of the Clementsian research group, see Joel Hagen, "Clementsian Ecologists: The Internal Dynamics of a Research Group," *Osiris* 8 (1992) (forthcoming).

53. Edward E. Dale, Jr., "Resolution of Respect: John Ernst Weaver 1884–1966," *Bulletin of the Ecological Society of America* 48 (1967): 107–109; Ronald C. Tobey, *Saving the Prairies: The Life Cycle of the Founding School of American Plant Ecology, 1895–1955* (Berkeley: University of California Press, 1981), chap. 7.

54. See CC for the extensive correspondence between the two men.

55. F. E. Clements to R. S. Woodward, April 25, 1918; F. E. Clements to J. E. Weaver, March 19, 1918, CC, Box 62.

56. See the correspondence between the two men during February and March 1918 and October and November 1918, CC, Box 62.

57. F. E. Clements to J. E. Weaver, December 23, 1918; Weaver to Clements, January 2, 1919; Weaver to Clements, January 7, 1919, CC, Box 62.

58. H. M. Hall to F. E. Clements, December 18, 1918, CC, Box 62.

59. H. M. Hall, "Heredity and Environment—As Illustrated by Transplant Studies," *Scientific Monthly* 35 (1932): 289–302; see also Hagen, "Experimentalists and Naturalists."

60. Clausen et al., *Experimental Studies.*

61. Hagen, "Clementsian Ecologists."

62. This argument is presented in its most extreme form in Daniel Simberloff, "A Succession of Paradigms in Ecology: Essentialism to Materialism and Probabilism," *Synthese* 43 (1980): 3–39. See also Robert H. Whittaker, "An Hypothesis Rejected: The Natural Distribution of Vegetation," in *Botany: An Ecological Approach,* ed. William A. Jensen and Frank B. Salisbury (Belmont: Wadsworth Publishing Co., 1972), 689–691; McIntosh, *Background of Ecology,* 80–83; and Bowers, *Sense of Place,* chap. 8.

63. Among prominent ecologists, Arthur Tansley, Henry Allan Gleason, Henry Chandler Cowles, William S. Cooper, Forrest Shreve, Burton Livingston, Arthur Vestal, Frank Egler, Stanley Cain, and Alex Watt publicly broke with Clements on key issues in plant ecology.

64. Garland E. Allen, "The Historical Development of the 'Time Law of Intersexuality' and its Philosophical Implications," in Leonie K. Piternik, ed., *Richard Goldschmidt: Controversial Geneticist and Creative Biologist* (Boston: Birkhauser Verlag, 1980).

65. James P. Collins, "Evolutionary Ecology and the Use of Natural Selection in Ecological Theory," *Journal of the History of Biology* 19 (1986): 257–288.

66. A. Hallam, *Great Geological Controversies* (Oxford: Oxford University Press, 1983), chap. 5; H. E. LeGrand, *Drifting Continents and Shifting Theories* (Cambridge: Cambridge University Press, 1988), chap. 4; Robert Muir Wood, *The Dark Side of the Earth* (London: George Allen & Unwin, 1985), chap. 4.

67. Ronald Good, *The Geography of the Flowering Plants* (New York: Longmans, Green and Co., 1947), pp. 50–52; Philip Stott, *Historical Plant Geography* (London: George Allen & Unwin, 1981), pp. 68–70; Joel B. Hagen, "Ecologists and Taxonomists: Divergent Traditions in Twentieth-Century Plant Geography," *Journal of the History of Biology* 19 (1986): 197–214.

68. Robert H. Whittaker, "A Consideration of Climax Theory: The Climax as a Population and Pattern," *Ecological Monographs* 23 (1953): 41–78, 53.

69. Hull, *Science as a Process,* 22–23.

4. The Metabolic Imperative

1. G. Evelyn Hutchinson, "Bio-Ecology," *Ecology* 21 (1940): 267–268.

2. Hutchinson's criticism was doubly unfair because Clements's coauthor, Victor Shelford, had discussed community metabolism on a number of occasions. The book referred to in the review is Clements and Shelford, *Bio-Ecology* (New York: John Wiley, 1939).

3. Charles Elton, *Animal Ecology* (London: Sidgwick & Jackson, 1927; London: Methuen, 1966), 1. The social and intellectual context of this scientific natural history is discussed by David Elliston Allen, *The Naturalist in Britain: A Social History* (London: Allen Lane, 1976), chap. 14. See also Sir Alister Hardy, "Charles Elton's Influence in Ecology," *Journal of Animal Ecology* 37 (1968): 3–8.

4. David L. Cox, *Charles Elton and the Emergence of Modern Ecology* (Ph.D. diss., Washington University, 1979), 5; Peter Crowcroft, *Elton's Ecologists: A History of the Bureau of Animal Population* (Chicago: University of Chicago Press, 1991), chap. 1.

5. Cox, *Charles Elton,* chap. 1; Crowcroft, *Elton's Ecologists,* chap. 1; Hardy, "Elton's Influence."

6. Elton, *Animal Ecology,* 56.

7. Ibid., 17, 52.

8. Ibid., 63–68.

9. A short history of this concept can be found in G. Evelyn Hutchinson, *An Introduction to Population Ecology* (New Haven: Yale University Press, 1978), chap. 5; see also Cox, *Charles Elton,* chap. 4.

10. Elton, *Animal Ecology,* 64.

11. Ibid.; the emphasis appears in the original text.

12. Richard C. Lewontin, "Adaptation," *Scientific American* 239 (September 1978): 212–230; R. H. Whittaker, S. A. Levin, and R. B. Root, "Niche, Habitat, and Ecotope," *American Naturalist* 107 (1973): 321–338.

13. Elton, *Animal Ecology*, 64.
14. Elton, *Animal Ecology*, 115; Stephen A. Forbes, "The Lake as a Microcosm," *Bulletin of the Illinois Natural History Survey* 15 (1925): 537–550 [originally published in the *Bulletin of the Peoria Scientific Association*, 1887].
15. Elton, *Animal Ecology*, 142–143.
16. Charles S. Elton, *The Pattern of Animal Communities* (London: Methuen, 1966).
17. Elton, *Animal Ecology*, 17; Charles Elton, *Animal Ecology and Evolution* (Oxford: Clarendon Press, 1930), 68.
18. Elton, *Animal Ecology*, 79. For a more detailed discussion of his views on the relationship between ecological theory and empirical research, see Elton, *Animal Ecology and Evolution*, chap. 3.
19. Elton, *Animal Ecology*, 98–99.
20. V. S. Summerhayes and C. S. Elton, "Contributions to the Ecology of Spitsbergen and Bear Island," *Journal of Ecology* 11 (1923): 214–286.
21. Ibid. See also Elton, *Animal Ecology*, 98–99, 103.
22. Elton, *Animal Ecology*, vi.
23. Ibid., 114.
24. Ibid., 113. Compare this with the similar example in A. M. Carr-Saunders, *The Population Problem: A Study in Human Evolution* (Oxford: Clarendon Press, 1922), 199–201. Elton acknowledged the importance this book, and the presentation of many of his ideas appears to have been influenced by reading it.
25. Elton, *Animal Ecology*, 115. For the classic account of the Kaibab deer, see D. Irvin Rasmussen, "Biotic Communities of Kaibab Plateau, Arizona," *Ecological Monographs* 11 (1941): 230–275.
26. Graeme Caughley, "Eruption of Ungulate Populations, with Emphasis on Himalayan Thar in New Zealand," *Ecology* 51 (1970): 53–72.
27. Elton, *Animal Ecology*, 144.
28. Elton, *Animal Ecology and Evolution*, 17. Elton's book contains the text of three lectures presented in 1929 at the University of London.
29. Ibid.
30. Daniel B. Botkin, *Discordant Harmonies: A New Ecology for the Twenty-first Century* (New York: Oxford University Press, 1990).
31. Hutchinson, *Population Ecology*, 214–215; see also G. E. Hutchinson, "The Lacustrine Microcosm Reconsidered," *American Scientist* 52 (1964): 334–341. For a discussion of this intellectual problem and its consequences, see Joel B. Hagen, "Research Perspectives and the Anomalous Status of Modern Ecology," *Biology and Philosophy* 4 (1989): 433–455.
32. For example, this intellectual problem is acknowledged by Eugene P. Odum, "Ecology: The State of Our Science," *Bulletin of the Ecological Society of America* 66, no. 1 (1985): 14–15.
33. Robert Graves and Alan Hodge, *The Long Week End: A Social History of Great Britain, 1918–1939* (New York: MacMillan, 1941), chap. 12; Charles Loch Mowatt, *Britain Between the Wars, 1918–1940* (Chicago: University of Chicago Press, 1955), chap. 4; Henry Pelling, *Modern Britain, 1885–1955* (New York: Norton, 1960), chaps. 5–6.
34. Mowatt, *Britain Between the Wars*, 260.
35. Cox, *Charles Elton*, chaps. 5–6; Crowcroft, *Elton's Ecologists*, chap. 2.
36. Elton, *Animal Ecology and Evolution*, 8.

37. For example, compare Elton's discussion of self-regulation in *Animal Ecology*, 144, with that in *Animal Ecology and Evolution*, 16–17.

38. For contrasting interpretations of Elton's commitment to group selection, see Cox, *Charles Elton*, 5; and William C. Kimler, "Advantage, Adaptiveness, and Evolutionary Ecology," *Journal of the History of Biology* 19 (1986): 215–233. The widespread use of various forms of group selection during this period is discussed in James P. Collins, "Evolutionary Ecology and the Use of Natural Selection in Ecological Theory," *Journal of the History of Biology* 19 (1986): 257–288.

39. Elton, *Animal Ecology and Evolution*, 9–10.

40. Ibid., 50. See also Charles Elton, "Theories of Evolution," *The School Science Review* 13 (1931): 55–61, 130–136.

41. Elton, "Theories of Evolution."

42. Elton, *Animal Ecology and Evolution*, 58; Elton, "Theories of Evolution," 133.

43. Elton, "Theories of Evolution," 133.

44. Carr-Saunders, *Population Problem*, chap. 19. Elton acknowledged Carr-Saunders, though not specifically for his ideas on tradition, in the preface of *Animal Ecology and Evolution*. For striking similarities in their discussions of tradition, compare Carr-Saunders, *Population Problem*, 414–415, and Elton, *Animal Ecology and Evolution*, 84–88.

45. Elton, *Animal Ecology and Evolution*, 90–91.

46. Elton, *Pattern of Communities*, 374–375.

47. A. MacFadyen, *Animal Ecology: Aims and Methods*, 2d ed. (London: Isaac Pitman, 1963), ix.

48. Hutchinson, *Population Ecology*, ix, 152; see also Cox, Charles Elton, chap. 3.

49. G. Evelyn Hutchinson, *The Kindly Fruits of the Earth: Recollections of an Embryo Ecologist* (New Haven: Yale University Press, 1979); see also Yvette H. Edmondson, "Some Components of the Hutchinson Legend," *Limnology and Oceanography* 16 (1971): 157–161.

50. Hutchinson, *Kindly Fruits*, 134.

51. Summerhayes and Elton, "Ecology of Spitsbergen," 231–233. A similar diagram, this time referred to as a food cycle, appears in Elton, *Animal Ecology*, 58.

52. Hutchinson, *Kindly Fruits*, 231; Kendall E. Bailes, *Science and Russian Culture in an Age of Revolutions: V. I. Vernadsky and His Scientific School, 1863–1945* (Bloomington: Indiana University Press, 1990), 198.

53. Hutchinson, *Kindly Fruits*, 233. Hutchinson also knew of Vernadsky's work through his colleague in the zoology department, Alexander Petrunkevitch, a Russian emigré who studied under Vernadsky.

54. G. Evelyn Hutchinson, "Circular Causal Systems in Ecology," *Annals of the New York Academy of Sciences* 50 (1948): 221–246, in particular, see 231, 236.

55. Edward S. Deevey, Jr., "Studies on Connecticut Lake Sediments. I. A Postglacial Climatic Chronology for Southern New England," *American Journal of Science* 237 (1939): 691–724; Deevey, "Studies on Connecticut Lake Sediments. III. The Biostratonomy of Linsley Pond," *American Journal of Science* 240 (1942): 233–264, 313–338.

56. G. E. Hutchinson and Anne Wollack, "Studies on Connecticut Lake Sediments II. Chemical Analyses of a Core from Linsley Pond, North Branford," *American Journal of Science* 238 (1940): 493–517.

57. Ibid., 509.
58. Ibid.
59. Deevey, "Studies on Lake Sediments. III," 248.
60. Hutchinson, *Kindly Fruits,* 223–224.
61. Deevey, "Studies on Lake Sediments. III," 235–236.
62. G. Evelyn Hutchinson, "Circular Causal Systems," 221–222. Note the similarity between this distinction and the later dichotomy between merological and holological approaches in Hutchinson, *Population Ecology,* 215.
63. Details of the early Macy conferences can be found in Walter David Hellman, "Norbert Wiener and the Growth of Negative Feedback in Scientific Explanation; with a Proposed Research Program of 'Cybernetic Analysis'" (Ph.D. diss., Oregon State University, 1982), chap. 6; and Steve Heims, "Encounter of Behavioral Sciences with New Machine-Organism Analogies in the 1940s," *Journal of the History of the Behavioral Sciences* 11 (1976): 368–373. For discussions of the impact of the Macy conferences on the development of modern biology, see Donna J. Haraway, "The High Cost of Information in Post–World War II Evolutionary Biology: Ergonomics, Semiotics, and the Sociobiology of Communication Systems," *Philosophical Forum* 13 (1981–82): 244–278; and Peter J. Taylor, "Technocratic Optimism, H. T. Odum, and the Partial Transformation of Ecological Metaphor After World War II," *Journal of the History of Biology* 21 (1988): 213–244.
64. Heims, "Behavioral Sciences." In an interview conducted on March 31, 1989, Hutchinson told me that the Macy conferences were among the most formative events in his career. Similar comments by Hutchinson can be found in Hellman, "Norbert Wiener and Negative Feedback," chap. 6. Howard Odum, however, who attended one conference with Hutchinson, thought it was a waste of time. Interview with the author, October 28, 1989.
65. Norbert Wiener, *Cybernetics or Control and Communication in the Animal and the Machine* (New York: John Wiley, 1948); Steve J. Heims, *John Von Neumann and Norbert Wiener: From Mathematics to the Technologies of Life and Death* (Cambridge: MIT Press, 1980); Hellman, "Norbert Wiener and Negative Feedback."
66. Steve Heims, "Gregory Bateson and the Mathematicians: From Interdisciplinary Interaction to Societal Functions," *Journal of the History of the Behavioral Sciences* 13 (1977): 141–159.
67. Sharon E. Kingsland, *Modeling Nature: Episodes in the History of Population Ecology* (Chicago: University of Chicago Press, 1985).
68. Hutchinson, "Circular Causal Systems," 237. See also Hutchinson and Wollack, "Studies on Connecticut Lake Sediments II," 509–511.
69. Wiener, *Cybernetics,* introduction; Hellman, "Norbert Wiener and Negative Feedback," chap. 4.
70. Ibid.
71. Kingsland, *Modeling Nature,* 106–116.
72. V. A. Kostitzin, "Evolution of the Atmosphere: Organic Circulation, Glacial Periods," reprinted in *The Golden Age of Theoretical Ecology: 1923–1940,* ed. Francesco M. Scudo and James R. Ziegler (New York: Springer-Verlag, 1978). The sketchy details of Kostitzin's life are presented by Scudo and Ziegler in "Vladimir Aleksandrovich Kostitzin and Theoretical Ecology," *Theoretical Population Biology* 10 (1976): 395–412.
73. Hutchinson, "Circular Causal Systems," 222–223.
74. Ibid., 230–231.

75. Ibid., 228.
76. Ibid., 242.
77. For example, see Gordon A. Riley, "Introduction," Yvette H. Edmondson, "Some Components of the Hutchinson Legend," and Alan J. Kohn, "Phylogeny and Biogeography of *Hutchinsonia*: G. E. Hutchinson's Influence through his Doctoral Students," in *Limnology and Oceanography* 16 (1971). This admiration is also evident in the many letters to Hutchinson from his former students and other young biologists who corresponded with him through the years. This correspondence fills several boxes in the G. Evelyn Hutchinson Papers, Sterling Library, Yale University.
78. Hutchinson, *Kindly Fruits*, 241.
79. Hutchinson, *Kindly Fruits*, 240–248, claims that his research was not appreciated by his colleagues at Yale. Nonetheless, he apparently had little difficulty offering courses in limnology, attracting bright students, and establishing a viable research program during his early years there.
80. Riley, "Introduction," 177.
81. Hutchinson, *Population Ecology*, 40.
82. G. E. Hutchinson to C. Sagan, September 13, 1973, Box 4, Numerical Sequence II, HP.
83. C. Juday to R. Pennak, 1942, quoted in Annamarie L. Beckel, "Breaking New Waters: A Century of Limnology at the University of Wisconsin," Special Issue: *Transactions of the Wisconsin Academy of Science, Arts and Letters* (1987), 24.

5. The Birth of a Specialty

1. Raymond L. Lindeman, "The Trophic-Dynamic Aspect of Ecology," *Ecology* 23 (1942): 399–418, 400.
2. Robert Edward Cook, "Raymond Lindeman and the Trophic-Dynamic Concept in Ecology," *Science* 198 (1977): 22–26.
3. A. G. Tansley, "The Use and Abuse of Vegetational Concepts and Terms," *Ecology* 16 (1935): 284–307.
4. Donald Worster, *Nature's Economy: A History of Ecological Ideas* (Cambridge: Cambridge University, 1977), 303.
5. Ronald C. Tobey, *Saving the Prairies: The Life Cycle of the Founding School of American Plant Ecology, 1895–1955* (Berkeley: University of California, 1981), 156. The differences between Clements's organismal concept and the later ecosystem concept are also strongly emphasized by Chunglin Kwa, "Representations of Nature Mediating Between Ecology and Science Policy: The Case of the International Biological Programme," *Social Studies of Science* 17 (1987): 413–442.
6. Tobey, *Saving the Prairies*, 8.
7. Sir Harry Godwin, "Sir Arthur Tansley: The Man and the Subject," *Journal of Ecology* 65 (1977): 1–26.
8. Letters between the two men are collected in the Department of Botany, Cambridge University (henceforth, TP), and in the Edith S. Clements and Frederic E. Clements Collection, American Heritage Center, University of Wyoming (henceforth, CC). Their face-to-face meetings were limited to a

very few occasions: the International Phytogeographical Excursions of 1911 and 1913, and a visit by Tansley to Clements's alpine laboratory in 1926.

9. Godwin, "Arthur Tansley."

10. Ibid., 23.

11. John Sheail, *Seventy-five Years In Ecology: The British Ecological Society* (Oxford: Blackwell, 1987), part 1; Godwin, "Arthur Tansley"; A. G. Tansley, "The Early History of Modern Plant Ecology in Britain," *Journal of Ecology* 35 (1947): 130–137; W. H. Pearsall, "The Development of Ecology in Britain," *Journal of Ecology* (supplement) 52 (1964): 1–12; Sir Edward Salisbury, "The Origin and Early Years of the British Ecological Society," *Journal of Ecology* (supplement) 52 (1964): 13–18.

12. Godwin, "Arthur Tansley."

13. F. F. Blackman and A. G. Tansley, "Ecology in its Physiological and Phytotopographical Aspects," *New Phytologist* 4 (1905): 199–203, 232–253. F. E. Clements to A. G. Tansley, July 30, 1905, TP.

14. Blackman and Tansley, "Ecology," 247.

15. Frederic Edward Clements, *Research Methods in Ecology* (Lincoln: University Publishing Co., 1905), 10; Joel B. Hagen, "Organism and Environment: Frederic Clements's Vision of a Unified Physiological Ecology," in *The American Development of Biology*, ed. Ronald Rainger, Keith R. Benson, and Jane Maienschein (Philadelphia: University of Pennsylvania Press, 1988).

16. F. F. Blackman to A. G. Tansley (undated), Tansley Papers, Botany Department, Cambridge University.

17. A. G. Tansley, "The Classification of Vegetation and the Concept of Development," *Journal of Ecology* 8 (1920): 118–149; A. G. Tansley, "Succession: The Concept and Its Values," *Proceedings of the International Congress of Plant Sciences* 1 (1926): 677–686; Tansley, "Use and Abuse."

18. F. O. Bower, "Botanical Bolshevism," *New Phytologist* 17 (1918): 106–107.

19. Tansley, "Classification of Vegetation," 126; Tansley, "Succession," 685–686.

20. Tansley, "Succession," 678–679.

21. John Phillips, "Succession, Development, the Climax, and the Complex Organism: An Analysis of Concepts," *Journal of Ecology* 22 (1934): 554–571; 23 (1935): 210–246, 488–508.

22. Ibid., 23 (1935) 505.

23. F. E. Clements to H. A. Spoehr, March 26, 1935, CC, Box 72.

24. Tansley, "Use and Abuse," 285.

25. Phillips, "Succession," 492.

26. J. C. Smuts, *Holism and Evolution* (New York: Macmillan, 1926).

27. Ibid., 339–340. Phillips initially accepted Smuts's claim that communities are organic without being true organisms, but by 1935 he had swung squarely behind Clements stronger organismal claim. Compare John Phillips, "The Biotic Community," *Journal of Ecology* 19 (1931): 1–24, 20; and Phillips, "Succession," 497.

28. Clements to H. A. Spoehr, March 26, 1935, CC, Box 72.

29. H. Levy, *The Universe of Science* (London: Watts, 1932), 77–81; Ralph Lyndal Worrall, *The Outlook of Science: Modern Materialism* (London: John Bale, 1933), 139–141; and Lancelot Hogben, *The Nature of Living Matter* (London: Kegan Paul, 1930), 289–301.

30. Tansley, "Use and Abuse," 297–298; emphasis in original.
31. Smuts, *Holism and Evolution,* 99.
32. Tobey, *Saving the Prairies,* 189–190.
33. See, for example, A. G. Tansley, *The Values of Science to Humanity* (London: Allen & Unwin, 1942). Tansley was one of a small group of vocal critics of the Marxist influence in British science; see Gary Werskey, *The Visible College* (London: Allen Lane, 1978), 280–285; and Tobey, *Saving the Prairies,* 187–190.
34. Frederic E. Clements and Ralph W. Chaney, *Environment and Life in the Great Plains* (Washington, D.C.: Carnegie Institution of Washington, 1936), 51–52; Tobey, *Saving the Prairies,* chap. 7; Worster, *Nature's Economy,* chap. 12.
35. Robert M. Crunden, *From Self to Society, 1919–1941* (Englewood Cliffs: Prentice Hall, 1972), x.
36. Tansley, "Use and Abuse," 299; Tansley, *Values of Science,* 30–31.
37. H. Godwin to M. Numata, October 8, 1979, TP. See also Harry Godwin, *Cambridge and Clare* (Cambridge: Cambridge University Press, 1985), chap. 17. In interviews that I conducted, Paul W. Richards and Jack L. Harley, biologists who knew Tansley during this period, also expressed skepticism toward Tobey's thesis.
38. A. G. Tansley, "British Ecology During the Past Quarter Century: The Plant Community and the Ecosystem," *Journal of Ecology* 27 (1939): 513–534. That Tansley did not totally abandon organicism is also evident in his continued use of organismal analogies in his discussion of human social activities; for example, see Tansley, *Values of Science,* 10.
39. H. Levy, *Universe of Science,* 45. For a biographical sketch of Levy, see Werskey, *Visible College,* chap. 2.
40. Levy, *Universe of Science,* 50–52.
41. Tansley, "Succession," 685–686.
42. Tansley, "Use and Abuse," 300.
43. G. E. Hutchinson and Anne Wollack, "Studies on Connecticut Lake Sediments II. Chemical Analyses of a Core From Linsley Pond, North Branford," *American Journal of Science* 238 (1940): 493–517, 510.
44. Donald Lawrence, quoted in Michael Finley, "Ray Lindeman: The 'Saint' of Cedar Creek," *Report* (June 1977): 2–3 [faculty newsletter of the University of Minnesota]. Lawrence was an instructor in botany at the University of Minnesota when Lindeman was a graduate student. A much more detailed account of Lindeman's personality can be found in Charles B. Reif, "Memories of Raymond Laurel Lindeman," *Bulletin of the Ecological Society of America* 67, no. 1 (1986): 20–25.
45. Application for Admission, June 21, 1932, Registrar's Office, Park College, Parkville, Missouri.
46. Reif, "Memories of Lindeman." G. Evelyn Hutchinson also emphasized these qualities in an interview, March 31, 1989.
47. His student notebooks are contained in Box 5 of the Raymond L. Lindeman Papers, Sterling Library, Yale University; henceforth, RLP.
48. For the importance of this group in shaping his ideas, see R. L. Lindeman to G. E. Hutchinson, February 21, 1941, RLP, Box 1.
49. Raymond L. Lindeman, "The Trophic-Dynamic Aspect." For a detailed account of the controversy surrounding Lindeman's paper, see Cook, "Raymond Lindeman."

50. Another striking example was the later collaboration between Hutchinson and Robert MacArthur, described in Sharon E. Kingsland, *Modeling Nature: Episodes in the History of Population Ecology* (Chicago: University of Chicago Press, 1985), chap. 8.

51. Raymond Laurel Lindeman, "Ecological Dynamics of a Senescent Lake" (Ph.D. diss., University of Minnesota, 1941). Four drafts of the trophic dynamic paper are contained in RLP, Box 2.

52. G. Evelyn Hutchinson, *Lecture Notes on Limnology* (New Haven: Osborn Zoological Laboratory, Yale University, privately distributed, 1941). This copyrighted manuscript is contained in the Hutchinson papers, Sterling Library, Yale University, henceforth HP. It is probably the same document referred to as *Recent Advances in Limnology* by Lindeman, "Trophic-Dynamic Aspect of Ecology."

53. Ibid., chap. 20. The mathematical notation found in Lindeman's trophic-dynamic paper was not used by Hutchinson in his 1941 lecture notes; however, the symbols were penciled into the margins of the manuscript. This is further evidence that Hutchinson's mathematical ideas were in flux during the period shortly before Lindeman arrived at Yale.

54. Lindeman transcript, Registrar's Office, Park College. In a letter, June 7, 1989, Charles B. Reif informed me that he and Lindeman audited a statistics course when they were students together at the University of Minnesota.

55. Raymond Laurel Lindeman, "Ecological Dynamics."

56. Lindeman, "Trophic-Dynamic Aspect," 399. The same point was made by Hutchinson, "Bio-ecology," *Ecology* 21 (1940): 267–268.

57. Lindeman, "Trophic-Dynamic Aspect," 400.

58. Ibid.

59. This distinction between matter and energy is clearly evident in the food-cycle diagram in Lindeman, "Trophic-Dynamic Aspect," 401. However, in his dissertation and an earlier paper, the same diagram appears without arrows representing energy flow; see Raymond L. Lindeman, "Seasonal Food-Cycle Dynamics in a Senescent Lake," *American Midland Naturalist* 26 (1941): 636–673, 637.

60. Lindeman, "Trophic Dynamic Aspect," 405.

61. R. L. Lindeman to J. R. Carpenter, February 26, 1942; Lindeman to W. S. Cooper, February 27, 1942, RLP, Box 1.

62. Howard T. Odum, "Primary Production in Flowing Waters," *Limnology and Oceanography* 1 (1956): 102–117; Eugene P. Odum, "The Strategy of Ecosystem Development," *Science* 164 (1969): 262–270.

63. See Lindeman's notes taken during Cooper's course on plant ecology, RLP, Box 5.

64. George L. Clarke, "Dynamics of Production in a Marine Area," *Ecological Monographs* 16 (1946): 323–335; A. MacFadyen, "The Meaning of Productivity in Biological Systems," *Journal of Animal Ecology* 17 (1948): 75–80; L. C. Birch and D. P. Clark, "Forest Soil as an Ecological Community With Special Reference to the Fauna," *Quarterly Review of Biology* 28 (1953): 13–36; Eugene P. Odum, *Fundamentals of Ecology* (Philadelphia: Saunders, 1953), chap. 3; Bernard C. Patten, "An Introduction to the Cybernetics of the Ecosystem: The Trophic-Dynamic Aspect," *Ecology* 40 (1959): 221–231; L. B. Slobodkin, "Energy in Animal Ecology," *Advances in Ecological Research* 1 (1962): 69–101; V. S. Ivlev, "The Biological Productivity of Waters," trans.

W. E. Ricker, *Journal of the Fisheries Research Board of Canada* 23 (1966): 1727–1759. Ivlev's paper was published in the Soviet Union in 1945. A mimeographed copy of Ricker's translation was widely circulated among North American ecologists shortly after World War II.

65. Cook, "Raymond Lindeman," provides a detailed account of this episode.

66. Cook, "Raymond Lindeman," claims that these were the reviewers. Given the stylistic similarities shared by the report of referee #1 and Juday's private correspondence concerning the Hutchinson school, there is little doubt as to the identity of this reviewer.

67. Comments of Referee #2, RLP, Box 1.

68. Comments of Referee #1, RLP, Box 1; emphasis in original.

69. T. Park to R. L. Lindeman, November 14, 1941, RLP, Box 1.

70. T. Park to R. L. Lindeman, March 23, 1942, RLP, Box 1.

71. Cook, "Raymond Lindeman."

72. See excerpts from Juday's correspondence, in Annamarie L. Beckel, "Breaking New Waters: A Century of Limnology at the University of Wisconsin," Special Issue: *Transactions of the Wisconsin Academy of Sciences, Arts and Letters* (1987).

73. Beckel, "Breaking New Waters."

74. Interview with Howard T. Odum, October 28, 1989; see also: G. Evelyn Hutchinson, *The Kindly Fruits of the Earth: Recollections of an Embryo Ecologist* (New Haven: Yale University Press, 1979), 248. W. T. Edmondson, another Hutchinson student, however, recalled a pleasant and productive stay at Juday's laboratory shortly before the Lindeman incident; letter to the author, March 22, 1989.

75. Charles Elton to G. Evelyn Hutchinson, April 21, 1943; NS 1, Box 2, HP; C. S. Elton, "An Ecological Textbook," *Journal of Animal Ecology* 23 (1954): 382–384.

76. R. L. Lindeman to W. S. Cooper, September 29, 1941, RLP, Box 1.

77. G. E. Hutchinson to T. Park, November 18, 1941, RLP, Box 1.

78. Eugene P. Odum, "Energy Flow in Ecosystems: A Historical Review," *American Zoologist* 8 (1968): 11–18; John Phillipson, *Ecological Energetics* (New York: St. Martin's Press, 1966), 15–17.

79. Edgar Nelson Transeau, "The Accumulation of Energy by Plants," *Ohio Journal of Science* 26 (1926): 1–10.

80. Ironically, Lindeman had forgotten about Thienemann's work, and Hutchinson had to remind him to look at it again. See G. E. Hutchinson to R. L. Lindeman, April 2, 1941; Lindeman to Hutchinson, April 7, 1941. In an early version of his paper Lindeman had attributed Thienemann's trophic concepts to Charles Elton. All items are in RLP, Box 1.

81. R. L. Lindeman to J. R. Carpenter, January 13, 1942, RLP, Box 1.

82. Emilio F. Moran, "Limitations and Advances in Ecosystem Research," in E. F. Moran, ed., *The Ecosystem Concept in Anthropology* (Boulder: Westview Press, 1984).

83. W. I. Vernadsky, "Problems of Biogeochemistry, II. The Fundamental Matter-Energy Differences Between the Living and the Inert Natural Bodies of the Biosphere," trans. George Vernadsky, *Transactions of the Connecticut Academy of Arts and Sciences* 35 (1944): 483–517, 487.

84. E. P. Odum, *Fundamentals of Ecology*, chaps. 1–2.

6. Ecology and the Atomic Age

1. Donald Worster, *Nature's Economy: A History of Ecological Ideas* (Cambridge: Cambridge University Press, 1985), 339.
2. Eugene P. Odum, *Fundamentals of Ecology,* 3rd ed. (Philadelphia: Saunders, 1971), 6, 467.
3. Eugene P. Odum, "Feedback Between Radiation Ecology and General Ecology," *Health Physics* 11 (1965): 1257–1262. See also Eugene P. Odum, "Ecology and the Atomic Age," *ASB Bulletin* 4, no. 2 (1957): 27–29.
4. Kenneth O. Emery, J. I. Tracey, Jr., and H. S. Ladd, "Geology of Bikini and Nearby Atolls," *Geological Survey Professional Paper* 260–A (Washington: U.S. Government Printing Office, 1954), 91. Eniwetok Atoll appears as little more than a dot on most maps of the Pacific Ocean. Geographical details of the atoll are provided by hydrographic charts #81523 (4th ed., 1976) and #81531 (3rd ed., 1984), Defense Mapping Agency Hydrographic Center, Washington, D.C.
5. Barton C. Hacker, *The Dragon's Tail: Radiation Safety in the Manhattan Project, 1942–1946* (Berkeley: University of California Press, 1987), 118, 160; Richard L. Miller, *Under the Cloud: The Decades of Nuclear Testing* (New York: Free Press, 1986), chap. 3; Richard Rhodes, *The Making of the Atomic Bomb* (New York: Simon and Schuster, 1986), epilogue.
6. Miller, *Under the Cloud,* 116; Rhodes, *Making the Atomic Bomb,* epilogue.
7. Miller, *Under the Cloud,* 306.
8. Neal O. Hines, *Proving Ground: An Account of the Radiobiological Studies in the Pacific* (Seattle: University of Washington Press, 1962), chap. 2; Roger Revelle, "Foreword," in Emery, Tracey, and Ladd, "Geology of Bikini," iii–vii. For a discussion of debates over the effects of radiation during this period, see George T. Mazuzan and J. Samuel Walker, *Controlling the Atom: The Beginnings of Nuclear Regulation, 1946–1962* (Berkeley: University of California Press, 1985).
9. Howard T. Odum and Eugene P. Odum, "Trophic Structure and Productivity of a Windward Coral Reef Community on Eniwetok Atoll," *Ecological Monographs* 25 (1955): 291–320; interview with Eugene P. Odum, December 12, 1988; interview with Howard T. Odum, October 28, 1989.
10. Odum and Odum, "Eniwetok Atoll," 291; emphasis in original.
11. Howard T. Odum, "Trophic Structure and Productivity of Silver Springs, Florida," *Ecological Monographs* 27 (1957): 55–112.
12. Eugene P. Odum, *Fundamentals of Ecology,* 2d ed. (Philadelphia: Saunders, 1959), vi; see also Eugene P. Odum, "The New Ecology," *BioScience* 14 (1964): 14–16.
13. Interview with E. P. Odum. The species were later identified by taxonomists at the Smithsonian Institution before the paper was submitted for publication.
14. Ibid.
15. Odum and Odum, "Eniwetok Atoll," 298.
16. For a detailed discussion of the diurnal flow method, see Howard T. Odum, "Primary Production in Flowing Waters," *Limnology and Oceanography* 1 (1956): 102–117; see also H. T. Odum, "Silver Springs."
17. Odum and Odum, "Eniwetok Atoll," 318.

18. George L. Clarke, "Dynamics of Production in a Marine Area," *Ecological Monographs* 16 (1946): 323–335.
19. Odum and Odum, "Eniwetok Atoll," 318; see also Peter J. Taylor, Technocratic Optimism, H. T. Odum and the Partial Transformation of Ecological Metaphor after World War II," *Journal of the History of Biology* 21 (1988): 212–244.
20. For example, see Eugene P. Odum, *Basic Ecology* (Philadelphia: Saunders, 1983), 5–6, 213–214, 508.
21. John M. Teal, "Community Metabolism in a Temperate Cold Spring," *Ecological Monographs* 27 (1957): 283–302. Teal discussed this early work with me during an interview, February 27, 1989.
22. Ibid., 283.
23. Ibid. See also John Phillipson, *Ecological Energetics* (New York: St. Martin's Press, 1966), 22.
24. The touchstone document in this historic episode is Vannevar Bush's 1945 report to the president, *Science: the Endless Frontier* (reprint; Washington, D.C.: National Science Foundation, 1960); see also Vannevar Bush, *Modern Arms and Free Men: A Discussion of the Role of Science in Preserving Democracy* (New York: Simon and Schuster, 1949). The origins of the postwar relationship between science and government is analyzed by Harvey M. Sapolsky, "Academic Science and the Military: The Years Since the Second World War," in *The Sciences in the American Context: New Perspectives,* ed. Nathan Reingold (Washington: Smithsonian Institution Press, 1979), and Sapolsky, *Science and the Navy: The History of the Office of Naval Research* (Princeton: Princeton University Press, 1990), chaps. 1–2.
25. Robert P. McIntosh, *The Background of Ecology: Concept and Theory* (Cambridge: Cambridge University Press, 1985), 205.
26. Eugene P. Odum, "Radiation Ecology at Oak Ridge," in *Environmental Sciences Laboratory Dedication,* ed. S. I. Auerbach and N. T. Milleman (Oak Ridge: Oak Ridge National Laboratory, 1980), 56.
27. Tobey Appel, "Science Managing for Uncle Sam: NSF Support for Biology, 1950–1963" (manuscript). See also Alan T. Waterman, "Federal Support of Fundamental Research in the Biological Sciences," *AIBS Bulletin* 1, no. 5 (1951): 11–17; Paul B. Pearson, "The Biological Program of the Atomic Energy Commission," *AIBS Bulletin* 3, no. 3 (1953): 17–19; and Orr E. Reynolds, "Support of the Biological Sciences by the Office of Naval Research" *AIBS Bulletin* 2, no. 2 (1952): 18–20.
28. Roger R. Revelle, "Observations on the Office of Naval Research and International Science, 1945–1960," an oral history conducted in 1984 by Sarah Sharp, Regional Oral History Office, Bancroft Library, University of California, Berkeley, 1986, 14. See also Sapolsky, *Science and the Navy.*
29. Richard G. Hewlett and Francis Duncan, *A History of the United States Atomic Energy Commission. Vol II. Atomic Shield, 1947–1952* (University Park: Pennsylvania State University Press, 1969), chap. 8; Richard G. Hewlett and Jack M. Holl, *Atoms for Peace and War, 1953–1961: Eisenhower and the Atomic Energy Commission* (Berkeley: University of California Press, 1989), chap. 9; Mazuzan and Walker, *Controlling the Atom,* chap. 1.
30. Hines, *Proving Ground,* chap. 1.
31. Mazuzan and Walker, *Controlling the Atom,* chap. 1; Nathaniel H. Stetson, "Early Government Considerations," in *The Savannah River and Its Environs: Proceedings of a Symposium in Honor of Ruth Patrick for 35 Years of Studies at*

the Savannah River, ed. John C. Corey (Aiken: Savannah River Laboratory, 1987).

32. Eugene P. Odum, "Early University of Georgia Research, 1952–1962," in *The Savannah River,* ed. Corey. Interview with Eugene Odum.

33. Odum, "University of Georgia Research."

34. Ibid., 45.

35. Ibid., 46.

36. Interview with John Teal. Mildred Teal and John Teal, *Portrait of an Island* (New York: Atheneum, 1964).

37. E. P. Odum, "New Ecology."

38. Joseph D. Laufersweiler, "John N. Wolfe: Joining the United States Atomic Energy Commission," *Bulletin of the Ecological Society of America* 71, no. 2 (1990): 225; Chunglin Kwa, "Mimicking Nature: The Development of Systems Ecology in the United States, 1950–1975" (Ph.D. diss., University of Amsterdam, 1989), 57.

39. Kwa, "Mimicking Nature," 62–67; E. P. Odum, "Ecology at Oak Ridge"; Alvin M. Weinberg, "Oak Ridge National Laboratory," *Science* 109 (1949): 245–248.

40. E. P. Odum, "Ecology at Oak Ridge"; Eugene P. Odum and Frank B. Golley, "Radioactive Tracers as an Aid to the Measurement of Energy Flow at the Population Level in Nature," in *Radioecology: Proceedings of the First National Symposium on Radioecology,* ed. Vincent Schultz and Alfred W. Klement, Jr. (New York: Reinhold Publishing, 1963); Eugene P. Odum and A. J. Pontin, "Population Density of the Underground Ant, *Lasius flavus,* as Determined by Tagging with ^{32}P," *Ecology* 42 (1961): 186–188; Edward S. Deevey, Jr., "Radiocarbon Dating," *Scientific American* 186, no. 2 (1952): 24–28.

41. For example, see George M. Woodwell, ed., *The Ecological Effects of Nuclear War* (Upton: Brookhaven National Laboratory, 1965); George M. Woodwell, "Effects of Ionizing Radiation on Terrestrial Ecosystems," *Science* 138 (1962): 572–577; Howard T. Odum and Robert F. Pigeon, eds., *A Tropical Rain Forest: A Study of Irradiation and Ecology at El Verde, Puerto Rico* (Washington, D.C.: Atomic Energy Commission, 1970).

42. This was the cautionary message in an article aimed at the general public. George M. Woodwell, "The Ecological Effects of Radiation," *Scientific American* 208, no. 6 (1963): 40–49.

43. E. P. Odum, *Fundamentals of Ecology,* 2d ed., 469.

44. Joseph S. Fruton, *Molecules and Life: Historical Essays on the Interplay of Chemistry and Biology* (New York: John Wiley, 1972), 456–469.

45. Interview with G. Evelyn Hutchinson, March 31, 1989; letter, W. T. Edmondson to author, March 22, 1989. The abortive attempt is mentioned in a footnote to a report of a successful experiment. See G. E. Hutchinson and V. T. Bowen, "A Direct Demonstration of the Phosphorus Cycle in a Small Lake," *Proceedings of the National Academy of Sciences* 33 (1947): 148–153.

46. Hewlett and Holl, *Atoms for Peace and War,* chap. 8; Pearson, "Biological Program."

47. E. P. Odum, *Fundamentals of Ecology,* 3rd ed., 93–96, 459–461.

48. Robert H. Whittaker, "Experiments with Radiophosphorus Tracer in Aquarium Microcosms," *Ecological Monographs* 31 (1961): 157–188.

49. Schultz and Klement, *Radioecology;* Daniel J. Nelson and Francis C. Evans, eds., *Symposium on Radioecology* (Ann Arbor, 1969); D. J. Nelson, ed., *Radionuclides in Ecosystems: Proceedings of the Third National Symposium on Radio-*

ecology (Oak Ridge: Oak Ridge National Laboratory, 1971). See also papers delivered at the Hanford Symposium on Radiation and Terrestrial Ecosystems published in *Health Physics* 11 (1965).

50. Spencer R. Weart, *Nuclear Fear: A History of Images* (Cambridge: Harvard University Press, 1988), chap. 8; Mazuzan and Walker, *Controlling the Atom,* chap. 2.

51. E. P. Odum, "University of Georgia Research."

52. H. J. Muller, "Race Poisoning by Radiation," *Saturday Review,* June 9, 1956, 9–11, 37–39; Mazuzan and Walker, *Controlling the Atom,* chap. 2; John Beatty, "Weighing the Risks: Stalemate in the Classical/Balance Controversy," *Journal of the History of Biology* 20 (1987): 289–319; Diane B. Paul, "'Our Load of Mutations' Revisited," *Journal of the History of Biology* 20 (1987): 321–335.

53. E. P. Odum, *Fundamentals of Ecology,* 2d ed., 467.

54. Rachel Carson, *Silent Spring* (Boston: Houghton Mifflin, 1962). Carson only briefly mentioned radiation effects in her book and not within the context of biological magnification.

55. E. P. Odum, *Fundamentals of Ecology,* 2d ed., 469.

56. Mazuzan and Walker, *Controlling the Atom,* chap. 12.

57. Ibid.

58. E. P. Odum, *Fundamentals of Ecology,* 2d ed., 481.

59. Ibid., 486. See also John N. Wolfe, "Impact of Atomic Energy on the Environment and Environmental Science," in *Radioecology,* ed. Schultz and Klement.

60. Mazuzan and Walker, *Controlling the Atom,* chap. 1.

61. Hines, *Proving Ground,* chap. 1.

62. Chandra Mukerji, *A Fragile Power: Scientists and the State* (Princeton: Princeton University Press, 1989).

63. For the scientific manpower shortage during the postwar period, see Alvin M. Weinberg, "The Federal Laboratories and Science Education," *Science* 136 (1962): 27–30; Hewlett and Holl, *Atoms for Peace and War,* chap. 9.

64. This was most strikingly true in genetics. See Beatty, "Weighing the Risks"; Mazuzan and Walker, *Controlling the Atom,* chap. 2.

65. Hewlett and Holl, *Atoms for Peace and War,* chaps. 8–9.

66. Daniel Simberloff, "A Succession of Paradigms in Ecology: Essentialism to Materialism and Probabilism," *Synthese* 43 (1980): 3–39.

67. That descriptive title was used by Stanley Auerbach; see the introduction to E. P. Odum, "Ecology at Oak Ridge."

68. Hewlett and Holl, *Atoms for Peace and War,* p. 253.

69. Ibid.

70. *Public Papers of the Presidents of the United States: Dwight D. Eisenhower, 1960–1961* (Washington, D.C.: U.S. Government Printing Office, 1961), 1035–1040.

71. E. P. Odum, "Ecology at Oak Ridge."

7. The New Ecology

1. Eugene P. Odum, "The New Ecology," *BioScience* 14, no. 7 (1964): 14–16.

2. Eugene P. Odum, *Fundamentals of Ecology* (Philadelphia: Saunders, 1953; 1959; 1971). Howard Odum wrote the important chapter on ecosystem energetics.

3. Robert L. Burgess, "United States," in *Handbook of Contemporary Developments in World Ecology,* ed. Edward J. Kormondy and J. Frank McCormick (Westport: Greenwood Press, 1981); Gordon H. Orians, "A Diversity of Textbooks: Ecology Comes of Age," *Science* 181 (1973): 1238–1239.

4. Interview with Eugene P. Odum, December 12, 1988.

5. Robert H. Whittaker, "An Hypothesis Rejected: The Natural Distribution of Vegetation," in *Botany: An Ecological Approach,* ed. William A. Jensen and Frank B. Salisbury (Belmont: Wadsworth, 1972), 689–691.

6. Eugene Pleasants Odum, "Variation in the Heart Rate of Birds: A Study in Physiological Ecology" (Ph.D. diss., University of Illinois, 1939).

7. Letter, E. P. Odum to the author, February 10, 1988.

8. Ibid. Correspondence between the Odums and Hutchinson can be found in the G. Evelyn Hutchinson papers, Sterling Library, Yale University; henceforth, HP.

9. Letter, E. P. Odum to the author.

10. Dianne Young, "Scholars and Brothers," *Southern Living* 23 (July 1988): 89–90; Peter J. Taylor, "Technocratic Optimism, H. T. Odum, and the Partial Transformation of Ecological Metaphor after World War II," *Journal of the History of Biology* 21 (1988): 213–244.

11. Howard Thomas Odum, "The Biogeochemistry of Strontium" (Ph.D. diss., Yale University, 1950). "A Minute of the Department Meeting of November 15, 1950," N.S. I, Box 6, HP. For the same reason, Odum had difficulty getting parts of the dissertation published; see H. T. Odum to Hutchinson, June 10, 1951, N.S. I, Box 6, HP.

12. In addition to the Mercer Award (1956), the two brothers received the Crafoord Prize, awarded by the Swedish Academy of Sciences, in 1987.

13. Taylor, "Technocratic Optimism"; interview with Howard T. Odum, October 28, 1989.

14. Interview with H. T. Odum.

15. Ibid.; H. W. Odum to G. E. Hutchinson, June 10, 1947, N.S. I, Box 6, HP.

16. Alfred J. Lotka, *Elements of Physical Biology* (Baltimore: Williams and Wilkins, 1925); the book was reprinted as *Elements of Mathematical Biology* (New York: Dover, 1956). For discussions of Lotka's intellectual program and its influence on ecology, see Sharon E. Kingsland, *Modeling Nature: Episodes in the History of Population Ecology* (Chicago: University of Chicago Press, 1985), chap. 2; Taylor, "Technocratic Optimism."

17. Ludwig von Bertalanffy, "The Theory of Open Systems in Physics and Biology," *Science* 111 (1950): 23–29; K. G. Denbigh, *The Thermodynamics of the Steady State* (London: Methuen, 1951); I. G. Prigogine, *Introduction to Thermodynamics of Irreversible Processes* (Springfield: Charles C. Thomas, 1955).

18. Taylor, "Technocratic Optimism."

19. David L. Hull, *Science as a Process: An Evolutionary Account of the Social and Conceptual Development of Science* (Chicago: University of Chicago Press, 1988), chap. 1.

20. Taylor, "Technocratic Optimism."

21. Raymond L. Lindeman, "The Trophic-Dynamic Aspect of Ecology,"

Ecology 23 (1942): 399–418; George L. Clarke, "Dynamics of Production in a Marine Area," *Ecological Monographs* 16 (1946): 323–335.

22. For example, see E. P. Odum, *Fundamentals of Ecology,* 2d ed., 25, 29, 146–147, 257; letter, E. P. Odum to the author.

23. Bernard C. Patten and Eugene P. Odum, "The Cybernetic Nature of Ecosystems," *American Naturalist* 118 (1981): 886–895; Eugene P. Odum, *Basic Ecology* (Philadelphia: Saunders, 1983), 45, 49.

24. E. P. Odum, *Fundamentals of Ecology,* 2d ed., 44. This sentence appears in the chapter on energetics written by Howard Odum. See also Howard T. Odum and Richard C. Pinkerton, "Time's Speed Regulator: The Optimum Efficiency for Maximum Power Output in Physical and Biological Systems," *American Scientist* 43 (1955): 331–343; Howard T. Odum, *Environment, Power, and Society* (New York: John Wiley, 1971).

25. H. T. Odum, *Environment, Power, and Society,* 97.

26. E. P. Odum, *Fundamentals of Ecology,* 2d ed., chap. 1.

27. Walter B. Cannon, *The Wisdom of the Body* (New York: Norton, 1932; rev. ed., 1939). Letter, E. P. Odum to the author.

28. Stephen J. Cross and William R. Albury, "Walter B. Cannon, L. J. Henderson, and the Organic Analogy," *Osiris* 3 (1987): 165–192.

29. E. P. Odum, *Fundamentals of Ecology,* 2d ed., 242.

30. Gregg Alden Mitman, "Evolution by Cooperation: Ecology, Ethics, and the Chicago School, 1910–1950" (Ph.D. diss., University of Wisconsin, 1988), chap. 5.

31. C. H. Waddington, *The Strategy of the Genes: A Discussion of Some Aspects of Theoretical Biology* (London: Allen & Unwin, 1957), chap. 2; I. Michael Lerner, *Genetic Homeostasis* (New York: John Wiley, 1954), 1–4; Th. Dobzhansky, "A Review of Some Fundamental Concepts and Problems of Population Genetics," *Cold Spring Harbor Symposia on Quantitative Biology* 20 (1955): 1–10; R. C. Lewontin, "The Adaptations of Populations to Varying Environments," *Cold Spring Harbor Symposia on Quantitative Biology* 22 (1957): 395–408; W. C. Allee, Orlando Park, Alfred E. Emerson, Thomas Park, and Karl Schmidt, *Principles of Animal Ecology* (Philadelphia: Saunders, 1949), 672; Lawrence B. Slobodkin, *Growth and Regulation of Animal Populations* (New York: Holt, Rinehart and Winston, 1961), chap. 11; V. C. Wynne-Edwards, *Animal Dispersion in Relation to Social Behaviour* (Edinburgh: Oliver and Boyd, 1962), chap. 1; David Lack, *Population Studies in Birds* (Oxford: Oxford University Press, 1966), 301.

32. Howard T. Odum and Eugene P. Odum, "Trophic Structure and Productivity of a Windward Coral Reef Community on Eniwetok Atoll," *Ecological Monographs* 25 (1955): 291–320; H. T. Odum and Pinkerton, "Time's Speed Regulator."

33. Odum and Odum, "Trophic Structure," 318.

34. Cannon, *Wisdom of the Body,* 316–320.

35. E. P. Odum, *Fundamentals of Ecology,* 3rd ed., 33–36; Patten and E. P. Odum, "Cybernetic Nature of Ecosystems."

36. Robert Lilienfeld, *The Rise of Systems Theory: An Ideological Analysis* (New York: John Wiley, 1978), introduction.

37. Ibid., introduction, chap. 8.

38. E. P. Odum, "New Ecology," 15; Robert P. McIntosh, *The Background of Ecology: Concept and Theory* (Cambridge: Cambridge University Press, 1985), 221.

39. McIntosh, *Background of Ecology*, 209; Chunglin Kwa, "Mimicking Nature: The Development of Systems Ecology in the United States, 1950–1975" (Ph.D. diss., University of Amsterdam, 1989), 42; George Van Dyne, "Ecosystems, Systems Ecology, and Systems Ecologists," in *Systems Ecology*, ed. H. H. Shugart and R. V. O'Niell (Stroudsburg: Dowden, Hutchinson & Ross, 1979). See also Kenneth E. F. Watt, "The Nature of Systems Analysis," in his *Systems Analysis in Ecology* (New York: Academic Press, 1966); R. L. Kitching, *Systems Ecology: An Introduction to Ecological Modelling* (St. Lucia: University of Queensland Press, 1983), chap. 1.

40. Eugene P. Odum, "Energy Flow in Ecosystems: A Historical Review," *American Zoologist* 8 (1968): 11–18, 16; E. P. Odum, "New Ecology."

41. E. P. Odum, "New Ecology," 15.

42. Ibid.

43. Bernard C. Patten, "Systems Ecology: A Course Sequence in Mathematical Ecology," *BioScience* 16 (1966): 593–598; George M. Van Dyne, "Systems Ecology: The State of the Art," in *Environmental Sciences Laboratory Dedication*, ed. S. I. Auerbach and N. T. Millemann (Oak Ridge: Oak Ridge National Laboratory, 1980).

44. Letter, Bernard Patten to the author, August 4, 1989; Kwa, "Mimicking Nature," chap. 2.

45. Letter, B. C. Patten to the author. The book was Henry Quastler, *Essays on the Use of Information Theory in Biology* (Urbana: University of Illinois Press, 1953).

46. Letter, B. Patten to the author.

47. Patten, "Systems Ecology"; Kwa, "Mimicking Nature," chap. 2. See also Kenneth E. F. Watt, "An Experimental Graduate Training Program in Biomathematics," *BioScience* 15 (1965): 777–780.

48. H. T. Odum, *Environment, Power, and Society*, chap. 9; Kwa, "Mimicking Nature," chap. 2.

49. Patten, "Systems Ecology," 596.

50. Watt, "Systems Analysis," 3.

51. Kitching, *Systems Ecology*, xix.

52. Howard T. Odum, "Ecological Potential and Analogue Circuits for the Ecosystem," *American Scientist* 48 (1960): 1–8; N. E. Armstrong and H. T. Odum, "Photoelectric Ecosystem," *Science* 143 (1964): 256–258.

53. H. T. Odum, *Environment, Power, and Society*, chap. 9.

54. Ibid., vii.

55. Ibid., 206.

56. Lilienfeld, *Rise of Systems Theory*, 227, chaps. 8–9. See also Taylor, "Technocratic Optimism"; Cross and Albury, "Walter Cannon."

57. Egbert G. Leigh, Jr., "The Energy Ethic," *Science* 172 (1971): 664–666.

58. Richard Levins and R. C. Lewontin, "Dialectics and Reductionism in Ecology," *Synthese* 43 (1980): 47–78; T.F.H. Allen and Thomas B. Starr, *Hierarchy: Perspectives on Ecological Complexity* (Chicago: University of Chicago Press, 1982), 40.

59. Patten and E. P. Odum, "Cybernetic Nature of Ecosystems."

60. Allen and Starr, *Hierarchy;* R. V. O'Neill, D. L. Angelis, J. B. Wade, and T.F.H. Allen, *A Hierarchical Concept of Ecosystems* (Princeton: Princeton University Press, 1986).

61. E. P. Odum, *Fundamentals of Ecology*, 2d ed., 7; emphasis in original. A similar, though less detailed, discussion appeared in the first edition.
62. Ibid.
63. H. T. Odum, *Environment, Power, and Society*, 9–11.
64. Ibid., 94–98. The "inventory of parts" phrase can be found in the index entry under natural history.
65. Donald Worster, *Nature's Economy: A History of Ecological Ideas* (Cambridge: Cambridge University Press, 1985), 313.
66. E. P. Odum, *Fundamentals of Ecology*, 1st ed., 317.
67. Van Dyne, "Ecosystems"; E. P. Odum, *Fundamentals of Ecology*, 1st ed., chaps. 12–14; H. T. Odum, *Environment, Power, and Society*, chap. 10.
68. E. P. Odum, *Fundamentals of Ecology*, 3rd ed., 6.
69. H. T. Odum, *Environment, Power, and Society*, chap. 10.
70. Ibid. George Woodwell, "Recycling Sewage Through Plant Communities," *American Scientist* 65 (1977): 556–562; Young, "Scholars and Brothers"; interview with E. P. Odum.
71. Taylor, "Technocratic Optimism."
72. H. T. Odum, *Environment, Power, and Society*, 274–276.
73. E. P. Odum, *Fundamentals of Ecology*, 3rd ed., 35.
74. W. I. Vernadsky, "The Biosphere and the Noösphere," *American Scientist* 33 (1945): 1–12.
75. Howard T. Odum, "The Stability of the World Strontium Cycle," *Science* 114 (1951): 407–411.
76. George T. Mazuzan and J. Samuel Walker, *Controlling the Atom: The Beginnings of Nuclear Regulation, 1946–1962* (Berkeley: University of California Press, 1985), chap. 9; Spencer R. Weart, *Nuclear Fear: A History of Images* (Cambridge: Harvard University Press, 1988), chap. 11.
77. George M. Woodwell, "Toxic Substances and Ecological Cycles," *Scientific American* 216, no. 3 (1967): 24–31.
78. Emilio F. Moran, "Limitations and Advances in Ecosystem Research," in his *The Ecosystem Concept in Anthropology* (Boulder: Westview Press, 1984).
79. John M. Teal, "Community Metabolism in a Temperate Cold Spring," *Ecological Monographs* 27 (1957): 283–302.
80. See Hutchinson's addendum to Lindeman, "Trophic-Dynamic Aspect."
81. Taylor, "Technocratic Optimism," 230.
82. Eugene P. Odum, "The Strategy of Ecosystem Development," *Science* 164 (1969): 262–270.
83. F. Herbert Bormann and Gene E. Likens, *Pattern and Process in a Forested Ecosystem: Disburbance, Development, and the Steady State Based on the Hubbard Brook Ecosystem Study* (New York: Springer-Verlag, 1979), 177; Stuart L. Pimm, *Food Webs* (New York: Chapman and Hall, 1982), 191–203. See also Eugene P. Odum, "Field Experimental Tests of Ecosystem-Level Hypotheses," *Trends in Ecology and Evolution* 5 (1990): 204–205.
84. E. P. Odum, "Strategy," 262.
85. Ibid., 266.
86. Odum and Odum, "Structure and Productivity," 291.
87. H. T. Odum, "Ecological Potential," 4.
88. Interview with H. T. Odum. Odum's ideas were criticized in Lawrence B. Slobodkin, "Ecological Energy Relationships at the Population

Level," *American Naturalist* 94 (1960): 213–236; Slobodkin, "Energy in Animal Ecology," *Advances in Ecological Research* 1 (1962): 69–101, 82. See also Taylor, "Technocratic Optimism."

89. Moran, "Ecosystem Research"; see also Eric Alden Smith, "Anthropology, Evolutionary Ecology and the Explanatory Limitations of the Ecosystem Concept," in *Ecosystem Concept*, ed. Moran.

90. Interview with E. P. Odum.

8. Evolutionary Heresies

1. Eugene P. Odum, "The New Ecology," *BioScience* 14, no. 7 (1964): 14–16. This attitude was expressed even more strongly by Ramon Margalef, who defined ecology as "the biology of ecosystems"; see Ramon Margalef, *Perspectives in Ecological Theory* (Chicago: University of Chicago Press, 1968), 4. See also Francis C. Evans, "Ecosystem as the Basic Unit in Ecology," *Science* 123 (1956): 1127–1128.

2. Lawrence B. Slobodkin, *Growth and Regulation of Animal Populations* (New York: Holt, Rinehart and Winston, 1961; New York: Dover, 1980), preface to the Dover ed.; Nelson G. Hairston, Sr., "This Week's Citation Classic," *Current Contents* (Agricultural, Biological, and Environmental Sciences), May 18, 1981, 20; Margalef, *Perspectives,* 4; Evans, "Ecosystem."

3. Robert L. Burgess, "The Ecological Society of America: Historical Data and Some Preliminary Analyses," in *History of American Ecology* (New York: Arno Press, 1977). A similar pattern of growth occurred in Great Britain; see Appendix, *Journal of Ecology* (supplement), 52 (1964): 244.

4. For example, one might compare the very broad research programs of prominent ecologists such as G. Evelyn Hutchinson and Robert H. Whittaker and the relatively specialized research of their many students; see Yvette H. Edmondson, "Some Components of the Hutchinson Legend," *Limnology and Oceanography* 16 (1971): 157–161; W. E. Westman and R. K. Peet, "Robert H. Whittaker (1920–1980): The Man and His Work," in *Plant Community Ecology: Papers in Honor of Robert H. Whittaker,* ed. R. K. Peet (Dordrecht: Dr. W. Junk, 1985).

5. For example, see the comments on evolution in Eugene P. Odum, *Fundamentals of Ecology,* 2d ed. (Philadelphia: Saunders Company, 1959), 217–221, 230, 242; Eugene P. Odum, "The Strategy of Ecosystem Development," *Science* 164 (1969): 262–270; Margalef, *Perspectives,* chap. 4.

6. Gordon H. Orians, "Natural Selection and Ecological Theory," *American Naturalist* 96 (1962): 257–263; J. L. Harper, "A Darwinian Approach to Plant Ecology," *Journal of Ecology* 55 (1967): 247–270. Two excellent analyses of this historical episode are provided by William C. Kimler, "Advantage, Adaptiveness, and Evolutionary Ecology," and James P. Collins, "Evolutionary Ecology and the use of Natural Selection in Ecological Theory," in *Journal of the History of Biology* 19 (1986). See also John L. Harper, "The Contributions of Terrestrial Plant Studies to the Development of Ecological Theory," in *Changing Scenes in the Natural Sciences, 1776–1976,* ed. Clyde E. Goulden (Philadelphia: Academy of Natural Sciences, 1977).

7. William B. Provine, *The Origins of Theoretical Population Genetics* (Chicago: University of Chicago Press, 1971), chap. 5; Ernst Mayr, *The Growth of*

Biological Thought: Diversity, Evolution, and Inheritance (Cambridge: Harvard University Press, 1982), chap. 12; Ernst Mayr and William B. Provine, eds., *The Evolutionary Synthesis: Perspectives on the Unification of Biology* (Cambridge: Harvard University Press, 1980).

8. Kimler, "Advantage," 231–232.

9. Stephen Jay Gould, "The Hardening of the Modern Synthesis," in *Dimensions of Darwinism: Themes and Counterthemes in Twentieth-Century Evolutionary Theory,* ed. Marjorie Grene (Cambridge: Cambridge University Press, 1983).

10. Kimler, "Advantage"; Collins, "Evolutionary Ecology."

11. V. C. Wynne-Edwards, *Animal Dispersion in Relation to Social Behaviour* (Edinburgh: Oliver and Boyd, 1962).

12. Kimler, "Advantage," 231.

13. R. C. Lewontin, "The Units of Selection," *Annual Review of Ecology and Systematics* 1 (1970): 1–18.

14. Sharon E. Kingsland, *Modeling Nature: Episodes in the History of Population Ecology* (Chicago: University of Chicago Press, 1985), chap. 8.

15. R. C. Lewontin, "Introduction," in *Population Biology and Evolution,* ed. Lewontin (Syracuse: Syracuse University Press, 1968).

16. R. C. Lewontin and J. L. Hubby, "A Molecular Approach to the Study of Genic Heterozygosity in Natural Populations. II. Amount of Variation and Degree of Heterozygosity in Natural Populations of *Drosophila psuedoobscura,*" *Genetics* 54 (1966): 595–609.

17. Collins, "Evolutionary Ecology."

18. Peter S. Dawson and Charles E. King, eds., *Readings in Population Biology* (Englewood Cliffs: Prentice-Hall, 1971), preface.

19. For example, compare E. P. Odum, *Fundamentals of Ecology,* 2d ed., 7, and Bernard C. Patten and Eugene P. Odum, "The Cybernetic Nature of Ecosystems," *American Naturalist* 118 (1981): 886–895, esp. 894.

20. For a review of the various arguments for holism and reductionism in ecology, see Robert P. McIntosh, *The Background of Ecology: Concept and Theory* (Cambridge: Cambridge University Press, 1985), 252–256.

21. Ernst Mayr, "Cause and Effect in Biology," *Science* 134 (1961): 1501–1506.

22. Collins, "Evolutionary Ecology."

23. Daniel Simberloff, "A Succession of Paradigms in Ecology: Essentialism to Materialism and Probabilism," *Synthese* 43 (1980): 3–39; J. Merritt Emlen, *Ecology: An Evolutionary Approach* (Reading: Addison-Wesley, 1973), 341. See also Emilio Moran, "Limitations and Advances in Ecosystem Research," and Eric Alden Smith, "Anthropology, Evolutionary Ecology, and the Explanatory Limitations of the Ecosystem Concept," in *The Ecosystem Concept in Anthropology,* ed. E. Moran (Boulder: Westview Press, 1984).

24. Although overstated, this is the important point made by Daniel B. Botkin, *Discordant Harmonies: A New Ecology for the Twenty-first Century* (New York: Oxford University Press, 1990).

25. For example, see E. P. Odum, *Fundamentals of Ecology,* 2d ed., 22–25.

26. Moran, "Limitations and Advances"; Smith, "Explanatory Limitations."

27. Kimler, "Advantage"; Collins, "Evolutionary Ecology."

28. David Lack, *Population Studies of Birds* (Oxford: Oxford University Press, 1966), 299–312.

29. Wynne-Edwards, *Animal Dispersion*, 4–9.
30. Ibid., 7.
31. Slobodkin, *Growth and Regulation*, 131. Although his argument for the evolution of prudent predation was similar to Wynne-Edwards's, Slobodkin denied that it was based upon group selection (see the footnote on page 181 of Slobodkin's book).
32. Wynne-Edwards, *Animal Dispersion*, 16, 129–131.
33. Ibid., 20.
34. Kimler, "Advantage."
35. Gregg Mitman, "From the Population to Society: The Cooperative Metaphors of W. C. Allee and A. E. Emerson," *Journal of the History of Biology* 21 (1988): 173–194.
36. Wynne-Edwards, *Animal Dispersion*, 6.
37. Ibid., 129.
38. Ibid., 131.
39. Ibid., 13–14.
40. Ibid., 9.
41. C. S. Elton, "Self-Regulation of Animal Populations," *Nature* 197 (1963): 634.
42. Lack, *Population Studies*, 311.
43. George C. Williams, *Adaptation and Natural Selection: A Critique of Some Current Evolutionary Thought* (Princeton: Princeton University Press, 1966), 3–4.
44. Ibid., 9.
45. Ibid., 16–19, 208–212.
46. Edward O. Wilson, *Sociobiology: The New Synthesis* (Cambridge: Harvard University Press, 1975), 120.
47. LaMont C. Cole, "Sketches of General and Comparative Demography," *Cold Spring Harbor Symposia on Quantitative Biology* 22 (1957): 1–15. See also Mitman, "Population to Society."
48. Williams, *Adaptation and Natural Selection*, 159.
49. Lewontin, "Levels of Selection."
50. Letter, M. J. Dunbar to the author, June 29, 1990.
51. Ibid.; M. J. Dunbar, "The Evolution of Stability in Marine Environments: Natural Selection at the Level of the Ecosystem," *American Naturalist* 94 (1960): 129–136. A later paper on the topic appeared in a festschrift for Hutchinson: M. J. Dunbar, "The Ecosystem as Unit of Natural Selection," *Transactions of the Connecticut Academy of Arts and Sciences* 44 (1972): 113–130.
52. Dunbar, "Ecosystem as Unit," 128.
53. Dunbar, "Evolution of Stability," 134.
54. Dunbar, *Ecological Development in Polar Regions: A Study in Evolution* (Englewood Cliffs: Prentice Hall, 1968), 100; Dunbar, "Ecosystem as Unit," 128.
55. Theological metaphors abound in the group selection literature. For example, note George Williams's reference to evolutionary "doctrine" (p. 4) and his sly allusion to John 14:6 in the last sentence of *Adaptation and Natural Selection*. Charles Elton accused Wynne-Edwards of presenting his arguments for group selection "pontifically"; see C. S. Elton, "Self-Regulation," 634. One of Dunbar's critics referred to his ideas as "biological theology," and Dunbar himself characterized his evolutionary writings as "heretical"; letter to the author. Similar comments appear in M. J. Dunbar, "Polar Marine Ecosystem

Evolution," in *The Arctic Seas: Climatology, Oceanography, Geology, and Biology*, ed. Yvonne Herman (New York: Van Nostrand Reinhold, 1989), 115.

56. Williams, *Adaptation and Natural Selection*, 247–250.

57. Eugene P. Odum, *Basic Ecology* (Philadelphia: Saunders, 1983), 485–486. This characterization of natural selection also appeared in all of the editions of Odum's *Fundamentals of Ecology*.

58. E. P. Odum, "Strategy."

59. Eugene P. Odum, *Fundamentals of Ecology*, 3rd ed. (Philadelphia: Saunders, 1971), 270–275; E. P. Odum, *Basic Ecology*, 483–487.

60. David Sloan Wilson, "Evolution on the Level of Communities," *Science* 192 (1976): 1358–1360; David Sloan Wilson, *The Natural Selection of Populations & Communities* (Menlo Park: Benjamin Cummings, 1980), chap. 1.

61. Simberloff, "Succession of Paradigms"; Moran, "Limitations and Advances"; Smith, "Explanatory Limitations."

62. For example, see Slobodkin, *Growth and Regulation*, 172.

63. Emlen, *Ecology*, 341; Douglas J. Futuyma, "Reflections on Reflections: Ecology and Evolutionary Biology," *Journal of the History of Biology* 19 (1986): 303–312.

64. Robert E. Ricklefs, *Ecology* (Portland: Chiron Press, 1973).

9. Big Ecology

1. W. Frank Blair, *Big Biology: The US/IBP* (Stroudsburg: Dowden, Hutchinson & Ross, 1977), chap. 6, estimates total federal expenditures at $40 to $50 million. The figure is set at $57 million by Philip M. Boffey, "International Biological Program: Was it Worth the Cost and Effort," *Science* 193 (1976): 866–868. See also Robert L. Burgess, "United States," in *Handbook of Contemporary Developments in World Ecology*, ed. Edward J. Kormondy and J. Frank McCormick (Westport: Greenwood Press, 1981); Robert P. McIntosh, *The Background of Ecology: Concept and Theory* (Cambridge: Cambridge University Press, 1985), 213–221.

2. I have borrowed this characterization from a lecture by Silvan Schweber, "Reflections on Physicists and Big Science After the Second World War" (Virginia Polytechnic Institute, Spring 1990).

3. Even observers as different in their political views and historical perspectives as the free-market capitalist Vannevar Bush and the Marxist J. D. Bernal agreed on this point; see Vannevar Bush, *Modern Arms and Free Men: A Discussion of the Role of Science in Preserving Democracy* (New York: Simon and Schuster, 1949), chap. 17; J. D. Bernal, *Science in History* (London: Watts, 1954), 578.

4. Richard A. Rettig, *Cancer Crusade: The Story of the National Cancer Act of 1971* (Princeton: Princeton University Press, 1977); see also "The Conquest of Cancer," *Science News* 98 no. 25 (1970): 459–460.

5. Howard T. Odum, *Environment, Power, and Society* (New York: John Wiley, 1971), 292–293.

6. According to Peter J. Taylor, Odum saw himself as a kind of ecological prophet; see "Technocratic Optimism, H. T. Odum, and the Partial Transformation of Ecological Metaphor after World War II," *Journal of the History of Biology* 21 (1988): 213–244. Taylor's claim finds support in a curious

cartoon of one of Odum's research projects in *Environment, Power, and Society,* 294–295.

7. Donald E. Stone, "The Organization for Tropical Studies (OTS): A Success Story in Graduate Training and Research," in *Tropical Rainforests: Diversity and Conservation,* ed. Frank Almeda and Catherine M. Pringle (San Francisco: California Academy of Sciences, 1988).

8. Howard T. Odum and Robert F. Pigeon, eds., *A Tropical Rain Forest: A Study of Irradiation and Ecology at El Verde, Puerto Rico* (Springfield: U.S. Atomic Energy Commission, 1970), ix–xi; B-12.

9. Ibid., chap. C-2. Similar irradiation studies were done at the Brookhaven National Laboratory, the Nevada Proving Grounds, and at Emory University. For a general discussion of the Brookhaven study, see George M. Woodwell, "The Ecological Effects of Radiation," *Scientific American* 208, no. 6 (1963): 40–49.

10. L. D. Gomez and J. M. Savage, "Searchers on that Rich Coast: Costa Rican Field Biology, 1400–1980," in *Costa Rican Natural History,* ed. Daniel H. Janzen (Chicago: University of Chicago Press, 1983).

11. Joseph D. Laufersweiler, "John N. Wolfe: Joining the United States Atomic Energy Commission" (paper presented at the 75th annual ESA meeting, 1990).

12. Buford R. Holt, book review of *A Tropical Rainforest,* in *Quarterly Review of Biology* 47 (1972): 113–114; see also Joseph Connell, "Project at El Verde," *Science* 172 (1971): 831–832.

13. Connell, "Project El Verde."

14. For a detailed critique of this aspect of the IBP, see: Chunglin Kwa, "Mimicking Nature: The Development of Systems Ecology in the United States, 1950–1975" (Ph.D. diss., University of Amsterdam, 1989).

15. E. B. Worthington, ed., *The Evolution of IBP* (Cambridge: Cambridge University Press, 1975).

16. Worthington, *Evolution of IBP,* chap. 1; Blair, *Big Biology,* chap. 1; W. H. Cook, "The IBP—Internationally and Nationally," in *Energy Flow—Its Biological Dimensions: A Summary of the IBP in Canada, 1964–1974,* ed. Thomas W. M. Cameron and L. W. Billingsley (Ottawa: Royal Society of Canada, 1975).

17. See the comments of C. H. Waddington and E. B. Worthington in Worthington, *Evolution of IBP,* 7, 53; Blair, *Big Biology,* 163.

18. Worthington, *Evolution of IBP,* 8–9.

19. Recollection of Josephine Doherty, in Chunglin Kwa, "Representations of Nature Mediating Between Ecology and Science Policy: The Case of the International Biological Programme," *Social Studies of Science* 17 (1987): 413–442.

20. Worthington, *Evolution of IBP,* 52; Eugene P. Odum, "The New Ecology," *BioScience* 14, no. 7 (1964): 14–16; Frederick E. Smith, "The International Biological Program and the Science of Ecology," *Proceedings of the National Academy of Sciences* 60, no. 1 (1968): 5–11.

21. Worthington, *Evolution of IBP,* 8–9; Kwa, "Representations of Nature."

22. See the letters of Nelson Hairston, LaMont Cole, and Thomas Park, reprinted in Blair, *Big Biology,* chap. 1; see also McIntosh, *Background of Ecology,* 213–221.

23. See the comments of Nelson Hairston in Blair, *Big Biology*, chap. 1; P.M.B., "'Boondoggle' Criticism Hits International Bio Program," *Science & Government Report* 2, no. 18 (1972): 1–3.

24. Blair, *Big Biology*, chap. 1.

25. Tobey A. Appel, "Science Managing for Uncle Sam: The National Science Foundation and Federal Support for Biology, 1950–1963" (manuscript).

26. See letter, LaMont Cole to W. F. Blair in Blair, *Big Biology*, 7–8.

27. Blair, *Big Biology*, 21; McIntosh, *Background of Ecology*, 215–216.

28. Worthington, *Evolution of IBP*, 10; for Revelle's self-assessment, see Roger Randall Dougan Revelle, "Observations on the Office of Naval Research and International Science, 1945–1960," an oral history conducted in 1984 by Sarah Sharp, Regional Oral History Office, The Bancroft Library, University of California, Berkeley, 1986, 22–27.

29. Blair, *Big Biology*, 22–23.

30. Kwa, "Representations of Nature"; Kwa, "Mimicking Nature." See also Robert Lilienfeld, *The Rise of Systems Theory: An Ideological Analysis* (New York: John Wiley, 1978), chap. 9.

31. For example, see the concluding sentence in Odum, "New Ecology."

32. Blair, *Big Biology*, chap. 3; compare this to the skillful use of testimony at congressional hearings by supporters of a National Cancer Institute described by Rettig, *Cancer Crusade*, chaps. 2, 7.

33. Blair, *Big Biology*, chap. 3; Kwa, "Representations of Nature."

34. Blair, *Big Biology*, 90.

35. The smaller programs were Origin and Structure of Ecosystems, Upwelling Ecosystems, Aerobiology, and Conservation of Ecosystems. These four projects accounted for less than 20 percent of the IBP budget for ecological research; see Burgess, "United States"; Blair, *Big Biology*, Appendix B.

36. Kwa, "Mimicking Nature," chap. 3; Worthington, *Evolution of IBP*, 67–70.

37. Kwa, "Mimicking Nature," chap. 3; Kwa, "Representations of Nature."

38. Burgess, "United States."

39. Blair, *Big Biology*, 92.

40. This was a major conclusion in a final report on the IBP presented to the National Science Foundation; see *Evaluation of Three of the Biome Studies Programs Funded Under the Foundation's International Biological Program (IBP)* (Columbus: Battelle Columbus Laboratories, 1975). A summary of this report can be found in Rodger Mitchell, Ramona A. Mayer, and Jerry Downhower, "An Evaluation of Three Biome Programs," *Science* 192 (1976): 859–865.

41. Kwa, "Mimicking Nature," chap. 3.

42. Blair, *Big Biology*, Appendix A, lists 203 participants in various data-gathering operations of the grassland biome study, excluding modelers. In addition, 135 graduate students were involved in the study. Several names, including some graduate students, appear more than once under different categories of work.

43. Ibid.

44. See comments of Frederick Smith in P.M.B., "Boondoggle Criticism." Opponents of the IBP had raised this type of criticism even before the biome studies had been proposed; see the letters of Nelson Hairston and Thomas Park in Blair, *Big Biology*, chap. 1. See also Kwa, "Mimicking Nature," chap. 3.

45. Worthington, *Evolution of IBP*, 67; Kwa, "Mimicking Nature."

46. R. L. Kitching, *Systems Ecology: An Introduction to Ecological Modelling* (St. Lucia: University of Queensland Press, 1983), 248; G. S. Innis, I. Noy-Meir, M. Godron, and G. M. Van Dyne, "Total System Simulation Models," in *Grasslands, Systems Analysis, and Man*, ed. A. I. Breymeyer and G. M. Van Dyne (Cambridge: Cambridge University Press, 1980), 774–775.

47. Daniel B. Botkin, "Bits, Bytes, and IBP," *BioScience* 27 (1977): 385.

48. The cartoon is reprinted in Worthington, *Evolution of IBP*, 70.

49. Blair, *Big Biology*, chap. 6; Worthington, *Evolution of IBP*, chap. 8; Orie L. Loucks, "The United States' IBP: An Ecosystem Perspective After Fifteen Years," in *Ecosystem Theory and Application*, ed. Nicholas Polunin (New York: John Wiley, 1986).

50. Ibid.

51. Burgess, "United States."

52. This concern was expressed by Nelson Hairston during the planning phase of the IBP, see Blair, *Big Biology*, 10–11. More recently, this criticism has been raised by Daniel Simberloff, "A Succession of Paradigms in Ecology: Essentialism to Materialism and Probabilism," *Synthese* 43 (1980): 3–39.

53. *Evaluation of Three Biome Studies*, 1–3; see also Boffey, "Worth the Cost."

54. For example, see the letters of W. Frank Blair and J. H. Gibson, *Science* 195 (1977): 822–823; see also Blair, *Big Biology*, chap. 6.

55. For example, the IBP is not even mentioned in discussions of biomes by Eric Pianka, *Evolutionary Ecology*, 3rd ed. (New York: Harper & Row, 1983), and Paul Colinvaux, *Ecology* (New York: John Wiley, 1986). The project is dismissed in one paragraph in Robert E. Ricklefs, *Ecology*, 3rd ed. (New York: Freeman, 1990), 260. Robert Leo Smith, *Ecology and Field Biology*, 4th ed. (New York: Harper & Row, 1990) briefly mentioned the IBP in a historical introduction and made scattered references to some biome projects in the body of the book. Eugene P. Odum, *Basic Ecology* (Philadelphia: Saunders, 1983) also mentioned the program, but most of his references to biome studies are pre-IBP. An exception, which provides a fairly detailed discussion of some IBP projects, is Charles J. Krebs, *Ecology: The Experimental Analysis of Distribution and Abundance*, 3rd ed. (New York: Harper & Row, 1985).

56. Burgess, "United States." For a discussion of the transition from the IBP to the Ecosystem Studies Program, see Blair, *Big Biology*, chap. 5.

57. For example, see Eric Alden Smith, "Anthropology, Evolutionary Ecology, and the Explanatory Limitations of the Ecosystem Concept," and Emilio F. Moran, "Limitations and Advances in Ecosystem Research," both of which appear in *The Ecosystem Concept in Anthropology*, ed. E. Moran (Boulder: Westview Press, 1984). See also Simberloff, "Succession of Paradigms."

58. P.M.B., "Boondoggle Criticism."

59. Simberloff, "Succession of Paradigms." A similar fear had been expressed by Thomas Park during the planning of the IBP; see Blair, *Big Biology*, chap. 1.

60. *Evaluation of Three Biome Studies.*

61. F. H. Bormann and Gene Likens, "Hydrologic-Mineral Cycle Interaction in a Small Watershed" (NSF Grant Proposal, 1963). I wish to thank Professor Bormann for providing me with a copy of this proposal. See also Gene E. Likens, F. Herbert Bormann, Robert S. Pierce, John S. Eaton, and Noye M. Johnson, *Biogeochemistry of a Forested Ecosystem* (New York: Springer-Verlag,

1977), v–ix; F. Herbert Bormann and Gene E. Likens, *Pattern and Process in a Forested Ecosystem: Disturbance, Development, and the Steady State Based on the Hubbard Brook Ecosystem Study* (New York: Springer-Verlag, 1979), chap. 1; McIntosh, *Background of Ecology,* 204–208. The funding history of the project is provided by *National Science Foundation Annual Reports,* 1963–1974. I wish to thank Tobey Appel for helping me to reconstruct the funding history of the Hubbard Brook project.

62. McIntosh, *Background of Ecology,* 205; Kwa, "Mimicking Nature," 76–77, presents Hubbard Brook as an imperfect version of the IBP biome projects.

63. Specifically, the hydrological and biogeochemical studies at Coweeta, North Carolina, one site in the Eastern Deciduous Forest Biome project, were quite similar to those done at Hubbard Brook.

64. Bormann and Likens, *Pattern and Process,* vii, 1; interview with F. H. Bormann, October 9, 1990.

65. Hubbard Brook was not officially part of the program, and at the end of the IBP Bormann and Likens refused to officially affiliate with the biome programs in the formation of a Committee on Ecosystem Studies; see Blair, *Big Biology,* 156; Kwa, "Mimicking Nature," 76–77.

66. F. H. Bormann, "Lessons from Hubbard Brook," in *Proceedings of the Chaparral Ecosystems Research Meeting,* ed. E. Keller, S. Cooper, and J. DeVries (Davis: California Water Resources Center, 1985); interview with F. H. Bormann.

67. Likens et al., *Biogeochemistry,* vi; interview with F. H. Bormann.

68. Bormann and Likens, *Pattern and Process,* chap. 1.

69. Ibid.

70. Ibid., 1–6.

71. Bormann, "Lessons from Hubbard Brook."

72. Eugene P. Odum, "The Strategy of Ecosystem Development," *Science* 164 (1969): 262–270.

73. Bormann and Likens, *Pattern and Process,* 177.

74. Ibid., 175.

75. Bormann, "Lessons from Hubbard Brook."

76. *Evaluation of Three Biome Studies,* I–4.

77. F. Herbert Bormann and Gene E. Likens, "The Nutrient Cycles of an Ecosystem," *Scientific American* 223, no. 4 (1970): 92–101; further information on publications and presentations can be found in an unpublished report, "Site Visit by the National Science Foundation to the Hubbard Brook Ecosystem Study" (October 1973). I am grateful to Professor Bormann for providing me with a copy of this document.

Epilogue

1. "Apollo's Return: Triumph Over Failure," *Time,* April 27, 1970, 12–14.

2. Howard T. Odum, *Environment, Power, and Society* (New York: John Wiley, 1971), 17–18, 285–288; Eugene P. Odum, *Ecology and Our Endangered Life-Support Systems* (Sunderland: Sinauer, 1989), 1–8.

3. Eugene P. Odum, *Basic Ecology* (Philadelphia: Saunders, 1983), 71.

4. E. P. Odum, *Life-Support Systems,* 62–63.
5. Ibid., 5.
6. J. E. Lovelock, *Gaia; A New Look at Life on Earth* (1979; Oxford: Oxford University Press, 1987).
7. Alfred C. Redfield, "The Biological Control of Chemical Factors in the Environment," *American Scientist* 46 (1958): 205–221.
8. For a brief summary of this research, see Lynn Margulis, *Early Life* (Boston: Science Books, 1982).
9. Richard A. Kerr, "No Longer Willful, Gaia Becomes Respectable," *Science* 240 (1988): 393–395.
10. E. P. Odum, *Basic Ecology,* 24–29; E. P. Odum, *Life-Support Systems,* 59–65.
11. Eugene P. Odum, "Ecology and the Atomic Age," *ASB Bulletin* 4, no. 2 (1957): 27–29. See also Eugene P. Odum, *Fundamentals of Ecology,* 2d ed. (Philadelphia: Saunders, 1959), ix.
12. Gordon H. Orians, "Diversity, Stability, and Maturity in Natural Ecosystems," in W. H. van Dobben and R. H. Lowe-McConell, *Unifying Concepts in Ecology* (The Hague: Dr. W. Junk, 1975).
13. J. Engelberg and L. L. Boyarsky, "The Noncybernetic Nature of Ecosystems," *American Naturalist* 114 (1979): 317–324. There were several responses to this article: S. J. McNaughton and Michael B. Coughenour, "The Cybernetic Nature of Ecosystems, *American Naturalist* 117 (1981): 985–990; Robert L. Knight and Dennis P. Swaney, "In Defense of Ecosystems," *American Naturalist* 117 (1981): 991–992; Bernard C. Patten and Eugene P. Odum, "The Cybernetic Nature of Ecosystems, *American Naturalist* 118 (1981): 886–895.
14. See William W. Murdoch's review of several papers presented at the 1990 meeting of the Ecological Society of America, "Equilibrium and Non-Equilibrium Paradigms, *Bulletin of the Ecological Society of America* 72, no. 1 (1991): 49–51.
15. Daniel B. Botkin, *Discordant Harmonies: A New Ecology for the Twenty-first Century* (New York: Oxford University Press, 1990), chap. 3; a standard historical work on this topic is Frank N. Egerton, "Changing Concepts of the Balance of Nature," *Quarterly Review of Biology* 48 (1973): 322–350.
16. S. A. Forbes, "On Some Interactions of Organisms," *Bulletin of the Illinois State Laboratory of Natural History* 1 (1880): 3–17, esp. 8.
17. David Sloan Wilson, "Evolution on the Level of Communities," *Science* 192 (1976): 1358–1360; Wilson, *The Natural Selection of Populations and Communities* (Menlo Park: Benjamin Cummings, 1980).
18. See the preface to the Dover reprint edition of Lawrence B. Slobodkin, *Growth and Regulation of Animal Populations* (New York: Dover, 1980). This provides an interesting contrast with Slobodkin's earlier optimism expressed on page 172.
19. G. Evelyn Hutchinson, *The Ecological Theater and the Evolutionary Play* (New Haven: Yale University Press, 1965).
20. Charles Darwin, *On the Origin of Species* [facsimile of the first edition, 1859] (Cambridge: Harvard University Press, 1964), 73.

Index